低醣飲食生活提案 **3**

Quick • Easy • Healthy

飽住落磅 輕鬆煮

健康瘦身餐

LOW-CARB MEAL PREP

陳倩揚 著

你得我得行動組團長

U0111022

萬里機構

推薦序一

由內靚到外的健康飲食態度

時間過得很快，距離上年書展已經一年。我在 2022 年聯同陳倩揚一起編寫《低醣飲食生活提案 2——全方位減脂營養天書》，分享正確體重控制的方法及打破很多營養飲食的謬誤。我們很高興見到讀者對低醣健康飲食產生不少興趣，非常踴躍地學習新的飲食模式，好讓身體變得輕盈又健康。

體重控制從來不容易，很多試過無數不同的方法，尤其是那些聲稱可以短期內大量減輕體重的極端飲食方法，到最後辛苦得來的最終結果竟然是體重回升。眼見很多人從此之後，不想再經歷那種減肥的痛苦。而當要從頭開始，亦覺得無比的困難。

從第一本的《健康輕鬆飽住瘦——低醣飲食生活提案》，陳倩揚用她的體重控制經驗和讀者分享一種可以持之以恆的健康飲食模式，而這本《低醣飲食生活提案 3——健康瘦身餐》，她再與大家分享更多她用心設計的簡單減醣食譜，教大家如何利用日常容易購買的完整食物，做出美味的低醣、高蛋白、高纖維，同時能提供不同抗氧化營養素的餸菜，使艱辛的體控旅程頓時變得自在。重點是：減肥不等

於要餓住進行，而是要吃得有飽足和幸福的感覺，才是一個持久的體重控制方法。

請大家繼續支持 Skye 的健康生活態度，由內靚到外！

林思為
顧問營養師

推薦序二

養 成 低 碳 低 卡 的 飲 食 生 活

要成功減重，每天都要消耗多於攝取的卡路里。消耗卡路里靠運動，而減少卡路里攝取則取決於飲食的分量和食物的選擇。不過，原來花功夫在控制飲食，是遠較運動更有效地達至每天負卡路里的效果。

每天消耗的卡路里由兩部分所組成——基礎代謝率和額外運動。基礎代謝率是指平時不運動時身體所消耗的卡路里，一個成年人每天的基礎代謝率大概是 1,100kcal 至 1,700kcal；另一個説法是一個成年人整天即使不做運動，單計基礎代謝率可消耗的能量為 1,500kcal。除非你是一個運動員或你經常做些極限且高能量的運動，運動在減重不是一個最重要的角色。比方説你每天跑步 30 分鐘，所消耗的熱量為 250kcal，但你只要吃一碗白飯已經補充了辛苦跑步所消耗的熱量。由此例子可見，由運動產生的消耗對整個減少卡路里工程其實幫助不大。最常見的情況是很多人在減重期間做了運動就放肆吃喝，導致體重有增無減，徒勞無功。基於這個原因，我們應該做好食物的選擇，尋找自己喜愛又好吃的餐單，持續低卡的生活。

明明減重就是計算卡路里收入和支出，卻時常聽到很多人

説已經把食物重量、卡路里進出計算得清清楚楚，但減重效果仍然強差人意。原因很簡單，當我們的體重開始下降，負責維持體重平衡的荷爾蒙就會迅速變化，增加食慾，減少運動意欲，務求阻止身體體重繼續下降。如果你的減重方式只是靠意志力進行，當身體遇上荷爾蒙變化，大腦釋出的訊號會令你減重的意志力及紀律性降低。

要應對這些荷爾蒙變化帶來的心態變化，我們要尋找自己喜愛又享受的減重生活方式，並持之以恆。除了多吃原形食物，也應該好好認識各款放進嘴巴的食物，知道每種食物的營養價值。最理想當然要學習健康的烹調方法，多在家中備餐、減少外食，讓低碳低卡的飲食變成生活的一種享受。自自然然，理想體重和健康生活也會隨之出現。

徐俊苗醫生
香港肥胖學會主席

Preface

目序

這個自序寫於 2023 年 2 月某天，當我快在一堆寫作功課中被淹沒，經一番作戰成功見岸後，呷一口咖啡，淡淡的榛子香氣讓我決定，將這份喜悅與成功感，換作轉場繼續努力寫的動力。換了氣氛與場景，打開手提電腦長期放置桌面的（方便隨時打開，想到就寫）「倩揚食譜」檔案，在一個一個食譜之前，開始寫這篇屬於《低醣飲食生活提案 3——健康瘦身餐》的自序。

對的，一年又一年過去，屬於《低醣飲食生活提案》系列，動筆之際已是來到第 3 本，綜合這兩年書展遇到過的每一位讀者，跟隨 Faceboook 專頁多年的 Followers，或是" My Friend Is You，My Pen Is You" 的 DM 系網友，再者就是走在街上或在街市遇見的你和妳，都會同聲問：
「倩揚，幾時會有純食譜呀？」
「睇完你的直播，好想有文字版隨時可以跟住做嘛！」
「我好傳統的，有本書睇住仲可以自己寫 Notes 呀！」
「你知唔知呀，你出本食譜書我方便周圍帶比啲 Friend 睇呀！」
「我要返工同埋湊仔湊図，無時間睇電話呀！」
（下省 500 個同類訴求 ^^）

好的，大家的意向我全單接收！在邁向轉字頭之齡前，又於適應再當學生之際，看着一堆功課、Projects、Mid-Term、考試時間線，澎湃地湧往每週日程的同時，我決定給自己多加一個目標——定於 2023 年 5 月前，挑戰一下自己時間管理的進步空間，騰出時間完成食譜書。當大家能夠看到這篇自序，即代表我的挑戰成功，完成了文字寫作，再完成了食譜製作及拍攝，趕上書商印刷的時間線，成功出版 Book 3 ！（掌聲掌聲⋯⋯）

大家常問，這份熱情何來？如何兼顧工作、家庭、經營社交平台、直播、拍片、剪片、出片，近來還加上繁重的功課，哪來的勁、哪來時間寫書？一想及如何回答，每每我都會瞬間落入內心的感性區，只是單純地覺得，人生太

短，年月匆匆可以不帶一點痕跡，需要好好把握擁有的時間，盡力去做能力做得的，去闖有信心去實踐的，僅此而已。更何況，盛載着你們每一位的支持與喜愛，聽着一個又一個成功減磅、成功重拾健康、成功重新能夠掌握自己人生信心的鼓舞分享，這一切全都印在心中，成為鼓勵自己不斷求進的力量。你們知道嗎？這些年來，維繫着我們的不只是幾本書、數個社交媒體，而是這份默默存在、環環相扣的熱誠、信心、鼓舞和滿滿的能量，在互相牽動，互相支持。

深情剖白後，除了藉此序再次感謝大家的支持外，當然希望這本食譜能夠成為每一位讀者家中的入廚天書之一！由新手入廚到煮理一家大小日常三餐，只要記着我在《低醣飲食生活提案 2——全方位減脂營養天書》不斷提到其中一個重點——「減醣餐不用分開煮」的大原則，為家中長者或成長中的小朋友稍微調整菜單，食譜內每個菜式都能夠幫助有需要 / 有興趣減磅的讀者做到——「開心食・輕鬆減・飽住瘦」。

插畫：陳羲穎 Audrick

倩揚 Facebook 專頁

你得我得行動組
Facebook 專頁

Contents

目錄

Chapter 1 　 Detox Water
維 他 命 排 毒 水

Chapter 2 　 Overnight Oats
一試愛上的「隔夜燕麥」

Chapter 3   Tsukemono

漬　　　　物

Chapter 4   Low-Carb Bento

日　常　減　醣　便　當

Chapter 5 **Low-Carb Sets**
日　常　減　醣　餐

Chapter 6   Hot Pot

打 邊 爐 飽 住 瘦

善用天然食材

　　當你實行低醣飲食法時，最重要是知道如何選擇食材和調味料，達到均衡營養攝取和美味的口感。重點是：揀得聰明，食得開心，才有動力和心情長久維持下去。以下是一些詳細建議：

1. 低醣飲食法建議減少碳水化合物攝取，但並不意味着必須食用清淡無味或單調的食物，只要注意避免過多的鹽和糖攝取，每餐加入適量的調味料絕對不是問題。天然的香草、香料、橄欖油、牛油果油及乳酪等都是不錯的選擇。另外，可使用醋、檸檬汁、芥末、洋葱、大蒜、辣椒和薑等天然調味料，來增加食物的風味。

2. 甜味的選擇：低醣飲食法建議減少糖的攝取，但不必完全放棄甜味。可以使用天然的甜味料，如蜂蜜、楓糖漿、無添加糖乾果類（如無花果及椰棗等），還可選用新鮮水果取代白糖。

3. 酸味的選擇：酸味料可以增加食物的鮮味，如檸檬汁、青檸汁、醋、柑橘類水果、番茄和酸梅等。這些天然的酸味料有助增加維他命 C 的攝取量；莓類亦含豐富的抗氧化物。

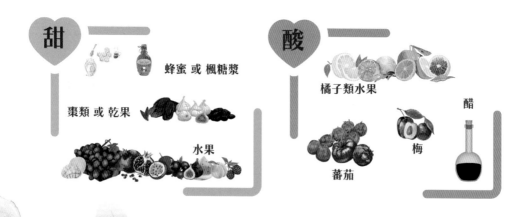

甜 | 蜂蜜 或 楓糖漿 | 棗類 或 乾果 | 水果

酸 | 橘子類水果 | 醋 | 梅 | 蕃茄

4. 苦味的選擇：有些蔬菜和水果帶苦味，如羽衣甘藍、火箭菜、芥菜、柚子、苦瓜、黑朱古力和茶葉等。這些苦味食材有助增加胃液分泌，促進消化和代謝。

5. 辣味的選擇：辣味食材可以增加食物的味道和風味，如辣椒、花椒、生薑及大蒜等，還有助促進身體代謝和消脂。

　　總的來説，善用天然食材調味食物，可以增加營養價值和減少身體的負擔，同時讓食物更加美味及健康。

我的常用食材表

我喜歡以天然食材炮製減醣美食,以下是我常用的食材及天然調味料,常備在家,隨手配搭,輕鬆煮出美味家常餐!

 常用乾貨

櫻花蝦 Sakura shrimps	營養豐富,含有豐富的鈣及磷,可強化骨骼,預防缺鈣所致的骨質疏鬆症。每 100 克櫻花蝦含 35 克蛋白質,適合高蛋白飲食。櫻花蝦含豐富蝦紅素,是一種強效抗氧化物,有益心臟健康。
蝦皮 Dried tiny shrimps	
蝦米 Dried shrimps	
黑白芝麻 Black and white sesames	含豐富鈣質、維他命 B、E、多元不飽和脂肪,對骨骼、神經系統、皮膚和心臟健康都有好處。可將芝麻撒在米飯或沙律,或作為甜點、蛋糕的健康食材。
紫菜碎 Dried seaweed	含豐富碘質,有助促進新陳代謝,保持甲狀腺健康。它是纖維含量高的低熱量食物,減肥期間可隨意使用。紫菜碎的天然鮮味有助減少用鹽的分量。
鰹魚碎 Dried bonito flakes	鰹魚的碳水化合物含量極低,蛋白質、DHA、鐵和維他命 B 含量高。烹調使用可增加風味及不添加額外的熱量。

蝦皮

蝦米

黑白芝麻

魚乾仔 Dried small fish	富含奧米加 3 脂肪酸、鈣和硒，有益心臟、骨骼和免疫力。約 45 克魚乾仔可提供 13 克蛋白質，而且碳水化合物含量非常低。
海帶芽 Wakame	含豐富的碘，對維持正常的甲狀腺和代謝功能非常重要。對不吃海鮮的人來說，它是極好的碘來源。海藻有助膳食中添加纖維，在減肥期間延緩飢餓感。
昆布 Kelp	天然增味劑，能為食物增加鮮味。含有對甲狀腺功能很重要的碘、鐵、鈣及微量礦物質。由於零脂肪，是減肥期間最有營養的食材之一。

魚乾仔

海帶芽

昆布

✓ 常用新鮮香草

羅勒 Basil	傳統用途包括治療蛇咬傷、感冒和鼻腔炎症。羅勒含有高濃度的丁香酚和檸檬鹼，具很強的抗氧化性，其提取物尤其是丁香酚可短暫降低高血壓。羅勒還含有豐富的葉黃素和玉米黃質，有益眼睛健康。
刁草 Dill	含強大的單萜類化合物，如檸檬烯、香芹酮和茴香醚，以及黃酮類化合物，還含大量維他命 A 和 C，以及微量葉酸、鐵和錳。這些植物化合物與降低患心臟病、中風和某些癌症風險有關。

羅勒

刁草

迷迭香

迷迭香 Rosemary	富含錳，對代謝健康最重要，還有助身體形成血塊及傷口癒合。迷迭香含鼠尾草酸，研究發現可減緩體內癌細胞生長，甚至降低患腫瘤的風險。迷迭香的鼠尾草酸和迷迭香酸具有強大的抗菌、抗病毒和抗真菌特性，定期食用有助降低感染風險。
百里香 Thyme	含有百里酚，可有益於呼吸系統健康，有助減少感冒。在老鼠的實驗研究，從百里香油提取的化合物有助保護胃壁免受潰瘍，增加胃部保護性胃黏液層，有助胃壁免受酸侵害。
薄荷葉 Mint	含有一種迷迭香酸的抗氧化劑和抗炎劑，改善牙齒和牙齦健康。薄荷含有薄荷糖，可能有助於分解痰液和黏液，有助化痰。薄荷有助增強食物的風味，減少鹽和糖的攝入量。
番茜 Parsley	熱量低，但富含維他命 A、K 和 C 等重要營養素。維他命 A 有益眼睛健康和免疫力；維他命 K 對骨骼有益；維他命 C 預防感冒和增強免疫力。
月桂葉 Bay leave	是維他命 A、維他命 B6 和維他命 C 的來源，可支持健康的免疫系統，並減輕炎症，幫助緩解哮喘症狀。此外，月桂葉可能有助減少肌肉痙攣，因含有 1,8-桉樹腦化合物，在小鼠研究證明可放鬆肌肉。
牛至 Oregano	含有香芹酚和百里酚精油，具有抗菌特性。在動物研究中，牛至提取物減少炎症，如自身免疫性關節炎、過敏性哮喘和類風濕性關節炎。牛至還有抗癌特性。
茴香 Fennel	茴香和茴香籽提供重要的營養成分，如維他命 C、鈣、鎂、鉀和錳，也含多酚抗氧化劑，是有效的抗炎劑，對健康有很強大的影響。

薄荷葉

番茜

月桂葉

✓ 常用乾香料

花椒 Sichuan peppercorns	對人體的消化系統有益，含鉀、銅、錳、磷、鋅、硒和鐵，是血紅蛋白的重要物質，將氧氣輸送到紅血管。
八角 Aniseed	有抗炎和抗菌特性，也具強大抗病毒能力，可能有助於降低患流感的風險。
薑黃 Turmeric	又稱薑黃素，常用於東南亞美食，以其抗炎症和抗氧化聞名，近年科學家建議多攝取薑黃素有助降低疾病風險，包括癌症、心臟病、糖尿病、關節炎和阿爾茨海默氏病等。
紅椒粉 Paprika	富含維他命 A、E、B6 和鐵，有益於眼睛及心臟健康，並有抗炎作用。由於富含鐵質，故有助降低患貧血的風險。
蒜粉 Garlic powder	大蒜粉是從大蒜提取，以「抗病毒」聞名。大蒜含有大蒜素，具抗衰老、抗炎和降血壓的功效。大蒜還有抗菌特性，是治療各種病毒、細菌、真菌和寄生蟲病的成分。
洋葱粉 Onion powder	洋葱含有花青素、硫槲皮素、硫化合物和硫代磺酸鹽，是強抗氧化劑，有助降低患心臟病和癌症的風險。
白胡椒粉 Ground white pepper	由成熟的去皮花椒製成，味道溫和，富含抗氧化劑和重要的微量營養素，包括錳、鐵和纖維。另富含胡椒鹼，可能有助減少炎症、降低血壓、保持消化系統功能。

花椒

八角

薑黃

紅椒粉

黑胡椒粉 Ground black pepper	由未成熟的胡椒植物核果製成，帶辛辣味，具有白胡椒相似的營養功能，除具有抗炎和降壓作用，還有抗癌和有助緩解腹瀉。
辣椒粉 Cayenne pepper	含辣椒素化合物，具辛辣味。它的類胡蘿蔔素含量特別高，對免疫力和眼睛健康都有好處。新鮮辣椒含有大量維他命 C，對膠原蛋白的產生、鐵質吸收和神經遞質合成很重要。

常用調味料

岩鹽 Halite	鈉含量略低於食鹽，富含鐵和其他礦物質，質地粗糙，比食鹽昂貴。
食鹽 Salt	是鹽類中鈉含量最高的一種，價錢較便宜，質地幼細，可能含有添加劑。
黃糖 Brown sugar **黑糖** Dark brown sugar	含有 95% 蔗糖和 5% 糖蜜，帶有拖肥糖風味和硬度，與白糖相比，營養價值很高。
楓糖漿 Maple syrup	由楓樹木質部汁液製成的糖漿，含有多酚、鈣、鉀、銅、鋅、錳、核黃素和硫胺素，具有抗炎和保護牙齦的功能。
龍舌蘭蜜 Agave syrup	從墨西哥龍舌蘭植物提取的糖，果糖含量高，比白糖甜 30%。升糖指數非常低，糖尿病患者可使用。
棕櫚糖 Palm sugar **椰糖** coconut sugar	由椰子樹汁液製成的調味糖，帶溫和的焦糖味。由於不像白糖般精製，含有礦物質。
蜜糖 Honey	果糖是蜜糖的主要糖分，其次是葡萄糖和蔗糖，其果糖含量令蜜糖比白糖甜些。
麻油 Sesame oil	具天然的堅果味，富含奧米加 6 脂肪酸、維他命 E 和 K。由於其揮發性很強，建議不會高溫烹調，多用於涼拌菜。

鰹魚粉或鰹魚汁 Katsuo Dashi or bonito sauce	含有天然的谷氨酸（glutamic acid），可帶出食物的鮮味，代替鹽使用有助減少鈉量，有助降低血壓。
大地魚粉 Dried flounder fish powder	大地魚是左口魚、比目魚或鰨沙魚曬乾製成。通常作為調味料，常用於煲湯煮菜，帶出鮮香味道。大地魚含豐富蛋白質及維他命 A、D、鈣、磷及鉀，脂肪含量低，更含豐富 DHA，有助智力發展。
沙薑粉 Sand ginger powder	沙薑研磨加工而成，色澤微黃，有濃郁香味，最常用作醃料。沙薑含有對人體呼吸系統和消化系統有益的桉葉油素，另具有抗炎和抗菌特性。
味噌 Miso	用鹽和麴菌發酵大豆製成，發酵過程促進益生菌生長，米麴霉是味噌中主要益生菌菌株，有益腸道健康、消化系統和炎症性腸病，有助建立良好的腸道微生態，維持免疫力。
菇粉 Mushroom powder	由乾香菇、牛肝菌、白樺茸及蘑菇製成。菇類含多醣體，具有調節免疫系統作用，可增強免疫力。蘑菇粉的抗氧化劑有助對抗自由基損傷，具抗衰老作用。

岩鹽

黃糖

大地魚粉

味噌

✓ 常用煮食油及沙律油

橄欖油 Olive oil	富含對心臟有益的單元不飽和脂肪，分為特級初榨橄欖油、初榨橄欖油和普通橄欖油。愈是初榨的油，煙點愈低，營養愈豐富，只適宜低溫烹調。高溫烹調宜選普通橄欖油。初榨橄欖油常用於沙律。
牛油果油 Avocado oil	富含單元不飽和脂肪，特別是油酸，有助減少血液的壞膽固醇，另富含維他命 E，對皮膚和肝臟健康都有好處。牛油果油常用於沙律。
芥花籽油 Canola oil	飽和脂肪含量低，富含維他命 E 和 K，還含有 28% 多元不飽和脂肪，主要是奧米加 3 脂肪酸 ALA，可在體內轉化為 DHA，有益大腦和眼睛健康。

✓ 常用種籽類

杏仁 Almond	在所有堅果中，杏仁含最豐富維他命 E——是一種強抗氧化劑，有益於心臟健康和免疫力。
核桃 Walnut	含有豐富的奧米加 3 脂肪酸，是所有堅果中最豐富，對大腦發育非常有益。核桃也含豐富銅，有助熱量代謝和神經健康。
腰果 Cashew	是所有堅果中碳水化合物含量最高，30 克腰果提供 10 克碳水化合物。腰果含有多種對骨骼健康重要的營養素，包括蛋白質、維他命 K、鎂和錳。
松子仁 Pine nuts	含有益心臟健康的多元不飽和脂肪和單元不飽和脂肪，有助提高膽固醇水平。 松子仁含鎂質，能調節血壓。松子仁富含保護性抗氧化劑，包括對眼睛健康有益的類胡蘿蔔素、葉黃素和玉米黃質。
花生 Peanut	富含單元不飽和脂肪、植物蛋白和膳食纖維，有益心臟及腸道，有助體重控制。花生富含菸酸、葉酸、鎂、磷和鉀，多酚可能有助於記憶，減少皮質醇、焦慮和抑鬱症狀。

榛子 Hazelnut	維他命 E 含量特別高，防止脂質過氧化，增強免疫反應。維他命 E 具抗癌特性，有研究表明可促進細胞凋亡，從而令癌細胞死亡。榛子還富含錳和銅，有益骨骼和血液健康。
開心果 Pistachio	特別富含維他命 B1 和 B6，有助維持神經系統和新陳代謝健康。開心果含有類胡蘿蔔素、葉黃素、玉米黃質及花青素，這些抗氧化物具有顯著的抗炎作用。
巴西果 Brazilian nuts	強抗氧化劑硒含量是所有堅果中最高的，有助預防癌症。巴西果的維他命 E 和鎂含量，有助控制血壓。
碧根果 / 長壽果 Peacan	是鈣、鎂和鉀的良好來源，有助降低血壓。碧根果的大部分脂肪是單元不飽和脂肪，有助降低血液的低密度脂蛋白。其升糖指數很低，有助穩定血糖。碧根果還含有奧米加 3 脂肪酸，可通過減少炎症有助緩解關節炎疼痛。
夏威夏果仁 Macadamia	主要含有益心臟健康的單元不飽和脂肪，有助降低患心臟病和 2 型糖尿病的風險。夏威夷果仁的抗氧化劑和類黃酮有助對抗炎症和減少細胞損傷；其含有生育三烯酚，可能有助預防某類型癌症和腦部疾病。

松子仁

巴西果

碧根果 / 長壽果

常用種籽類

南瓜籽 Pumpkin seeds	是色氨酸的天然來源，有助促進睡眠；南瓜籽的鋅、銅和硒也影響睡眠時間和質量；鋅有助免疫系統對抗細菌和病毒。
大麻籽 Hemp seeds	含大量奧米加 3 脂肪酸，有利素食者獲取對身體和大腦有益的營養素。大麻籽含極少碳水化合物，膳食纖維很高，有助紓緩消化系統，維持良好的腸道微生態。
罌粟籽 Poppy seeds	含大量膳食脂肪，其中大部分是奧米加多元不飽和脂肪酸亞油酸；另富含植物蛋白、維他命 E 和錳。罌粟籽有助穩定血糖，幫助形成良好的腸道微生態。
葵花籽 Sunflower kernel	維他命 E 和硒含量特別高，有抗氧化的作用，保護身體細胞免受自由基傷害。葵花籽有助降低血壓、膽固醇和血糖。
紫蘇籽 Perilla seeds	紫蘇籽油是必需脂肪酸的豐富來源，如 α-亞麻酸和亞油酸。紫蘇籽具有抗癌、抗糖尿病、抗哮喘、抗菌、抗炎、抗氧化和保護心臟的作用。
亞麻籽 Flaxseed	膳食纖維含量很高，有助緩解便秘，而且含豐富奧米加 3 脂肪酸，有助降低低密度脂蛋白，降低患心臟病的風險。亞麻籽含最豐富的木脂素——是植物雌激素的天然來源，有助降低血壓和動脈炎症。
奇亞籽 Chia seeds	28 克奇亞籽提供 11 克膳食纖維，有益腸道和心臟健康。奇亞籽富含奧米加 3 脂肪酸，蛋白質含量高，有助體重管理過程中增加飽腹感。奇亞籽的水溶性纖維更有助穩定血糖。

南瓜籽

葵花籽

奇亞籽

常備罐頭豆（無添加調味）

黑豆 Black beans	含有抗性澱粉和豐富膳食纖維，有助增加飽腹感和穩定血糖。其脂肪含量很低，蛋白質含量很高，葉酸和鐵含量特別高，非常適合孕婦食用。
紅腰豆 Red kidney beans	最常用的豆類，富含纖維和蛋白質，有助降低低密度脂蛋白，增強肌肉質量。與所有豆類一樣，紅腰豆的升糖指數較低，有助控制血糖水平。
鷹咀豆 Chickpeas	豐富的膳食蛋白質和碳水化合物，脂肪含量很少，提供豆類中最高的蛋白質來源。其葉酸含量特別高，是人體必需營養素，是懷孕和預防神經管缺陷的重要營養素。
利馬豆 White beans/ Butter beans	主要由複合碳水化合物、優質植物蛋白和極少量脂肪組成。利馬豆含有大量鐵質，預防貧血，而且可控制體重、緩解便秘和降低壞膽固醇水平。
蠶豆 Fava beans	富含礦物質錳，維持骨骼結構和密度。從飲食攝取足夠錳質，有助預防骨質疏鬆症。蠶豆富含左旋多巴（L-dopa），身體將這種化合物轉化為神經遞質多巴胺，研究表明可能降低患柏金遜病的風險。

黑豆

紅腰豆

鷹咀豆

小米 Millet	不含麩質的穀物，適合對麩質敏感或過敏人士食用。研究表明與其他穀物相比，小米有助身體將壞膽固醇降低 1-2 %。
藜麥 Quinoa	與其他全穀物相比，藜麥含最高的蛋白質，100 克藜麥提供 4 克蛋白質。藜麥是抗氧化劑和礦物質的良好來源，比其他普通穀物提供更多鎂、鐵、纖維和鋅。藜麥是無麩質穀物，適合對麩質不耐受人士。
糙米（玄米） Brown rice	只去除外殼，保留營養豐富的麩皮和胚芽，糙米含有白米缺乏的營養成分——維他命、礦物質和抗氧化劑。糙米的蛋白質含量比白米高，有助增加飽腹感和穩定血糖。另外，錳含量特別高，有助骨骼發育、傷口癒合、肌肉收縮代謝和神經功能。
紅米 Red rice	纖維和蛋白質含量略高於白米，富含類黃酮抗氧化劑、花青素、芹菜素、楊梅素和槲皮素，具很強的抗氧化作用。它還富含鐵，有利維持血紅蛋白水平。
黑米或紫米 Black rice	黑紫色來自花青素色素，具強大的抗氧化特性。黑米含有比糙米更高的蛋白質，100 克黑米提供 9 克蛋白質。研究表明，黑米有超過 23 種抗氧化劑，是所有穀物中抗氧化活性最高的。
原片大燕麥 Rolled oats	富含 β-葡聚醣，可降低膽固醇。每天攝入 70 克燕麥有助降低壞膽固醇 7-10%。燕麥片富含蛋白質和纖維，有助穩定血糖水平。

小米

藜麥

原片大燕麥

糙米（玄米）

紅米

黑米或紫米

天然甜味食材

紅棗 Red date	富含多種抗氧化化合物，主要是類黃酮、多醣和三萜酸，維他命 C 含量也高。廣泛用於改善睡眠質量和大腦功能。曬乾後常用於中式糖果和甜點。
椰棗 Coconut date	富含鐵、鉀、、銅、鎂、鈣和維他命 B，帶濃郁的焦糖甜味，常替代甜點糖分；但其熱量很高，宜適量食用。
杏甫乾 Preserved apricot	含 β 胡蘿蔔素、葉黃素和玉米黃質等抗氧化劑，有助對抗體內的自由基。其主要黃酮類化合物有助減少身體炎症；維他命 A 和 E 也有利眼睛健康。
藍莓乾 Dried blueberry	含豐富的花青素及維他命 K，有利促進血液健康。由於藍莓是低碳水化合物水果，升糖指數低，是糖尿病者的理想食物。
紅莓乾 Dried raspberry	富含生物活性植物化合物和抗氧化劑，所含的 A 型原花青素可防止大腸桿菌附在膀胱和泌尿道內壁，有助降低泌尿道感染的風險，並可防止幽門螺旋桿菌附在胃壁，降低胃癌的風險。
杞子 Chinese wolfberry	是超級食品，含大量鐵、葉黃素和玉米黃質。葉黃素和玉米黃質與保護眼睛和降低眼疾風險有關（與年齡相關性的黃斑變性）。杞子的類胡蘿蔔素和其他酚類化合物可提高免疫力、預防腫瘤和保持大腦健康。

紅棗

椰棗

桂圓肉 Dried longan	含有豐富的維他命 C、鉀和鐵質,有增強免疫力、降血壓、減少貧血等作用;但其升糖指數可能較高,糖尿病者慎用。
提子乾 Raisin	含大量鐵、銅和維他命,製造紅細胞和輸送氧氣到全身,有助預防貧血。所含的多酚,可保護眼睛細胞免受自由基傷害,保護眼睛免受侵害,如黃斑變性和白內障。
無花果 Dried fig	富含大量維他命 C 和 K、磷、鉀、鈣、鎂、纖維和強大的抗氧化劑。雖含糖量高,但這種天然甜味有助健康及減輕體重。其高鉀和鈣含量特別有利控制血壓和骨骼健康。

桂圓肉

提子乾

無花果

常用芝士

茅屋芝士 Cottage cheese	是一種柔軟的白色芝士，味道溫和。與其他芝士相比含較少卡路里，100 克茅屋芝士提供約 84 千卡、11 克蛋白質、4 克碳水化合物和 2 克脂肪。茅屋芝士含豐富酪蛋白（casesin），有助增加飽腹感。
羊奶芝士 Feta cheese	用綿羊奶或山羊奶製成，與其他芝士相比，其鈉和脂肪含量較高。100 克羊奶芝士提供約 14 克蛋白質和 493 毫克鈣，建議控制食用量。
莫札瑞拉芝士 Mozzerella cheese	柔軟溫和的新鮮芝士，起源於意大利。100 克莫札瑞拉芝士提供 22 克蛋白質、693 毫克鈣和大量維他命 B_{12}。相同分量的脫脂莫札瑞拉芝士最多提供 32 克蛋白質。
車打芝士 Cheddar cheese	是一種淡黃色、中等硬度的牛奶芝士，陳年車打芝士有濃烈和辛辣味，質地易碎，其熱量、脂肪和鈉含量相對較高。100 克車打芝士提供 23 克蛋白質、720 毫克鈣（幾乎相等兩杯牛奶）和極低糖分。
布里芝士 Brie	具忌廉般的質地，香氣獨特，味道溫和，經常與麵包、餅乾或水果一起食用。布里芝士含 30% 脂肪、20% 蛋白質和極低碳水化合物，還含豐富的核黃素、維他命 B_{12} 和鈣。

莫札瑞拉芝士

布里芝士

特別鳴謝：
食材營養資訊由顧問營養師林思為提供

日常食材對身體的重要性

順帶一提……

大家記得從日常飲食中均衡攝取充足的微量元素，這對維持正常的生理功能和健康起了關鍵性的作用，攝取不足可能會產生以下問題：

1. 營養不均衡：影響身體的正常功能和代謝過程，並增加患病風險。

2. 免疫功能下降：缺乏這些微量元素，可能會導致免疫功能下降，使身體更容易受到感染和疾病的侵害。

3. 骨骼問題：攝取不足可能增加骨質疏鬆和骨折的風險。

4. 能量代謝問題：可能影響能量的正常生成和利用，導致疲勞和代謝問題。

以下是其中幾款礦物質的注意重點，大家可以從中參考。其實只要不挑吃，多吃不同顏色的食物，做到均衡攝取並不困難。

Magnesium 鎂

攝取不足：疲勞、頭痛頭暈、抽筋、記憶力下降、注意力不集中、焦慮、憂鬱、生理期不適

富含鎂之食物：
豆類、果仁類、三文魚、黑朱古力、菠菜、胡蘿蔔、番茄、羽衣甘藍、芹菜、蘆筍、蘋果、牛油果、莓類、亞麻籽、薑黃、藻類等

Selenium 硒

攝取不足：易脫髮、免疫調節、抗氧化能力下降、憂慮、情緒低落

富含硒之食物：
豆腐、雞蛋、燕麥、杏仁、菇類、菠菜、西蘭花、大白菜、胡蘿蔔、番茄、蘋果、士多啤梨、螺旋藻等

Zinc 鋅

攝取不足：易脫髮、傷口癒合差、緊張、疲倦

富含鋅之食物：
甜菜根、菠菜、西蘭花、胡蘿蔔、番茄、羽衣甘藍、芹菜、蘆筍、蘋果、莓類、亞麻籽、薑黃、螺旋藻等

維他命B12

必須從飲食中
攝取維他命B12

攝取不足：注意力、集中力下降、焦躁、易出現貧血症狀、代謝力下降、易疲倦

富含維他命B12之食物：
肉類魚類、乳製品、植物奶、糙米、泡菜、蜆類、紅菜頭、大麥草、藻類、姬松茸等

維他命B12參與碳水化合物、蛋白質和脂肪的代謝。
蛋白質的代謝和吸收，尤其需要維他命B12才能順利進行

現成醬汁的陷阱

　　打開雪櫃，大家齊齊數數究竟收藏了多少款各式各樣的現成調味醬汁？

　　看文章前，不如順道檢視一下各醬汁的賞味期限及到期日，邊打點邊清理一下存貨。

　　是的，現成的醬料的確非常吸引，除方便之外，還好像隱藏了一絲味道保證的安全感，煎炒煮燜燉焗時加一點，「難食極有個譜啩！」我彷彿看到了大家邊看邊點頭的樣子。

　　記得年輕時初入廚，看食譜書、看煮食 Blog 學煮不同菜式，不知不覺間囤積了各式各樣、大樽小樽的醬料，直至搬家清理雪櫃，猛然發現原來很多醬料只用了數次，或長時間沒煮那菜式而過期。由那次搬家開始，加上因工作關係與不同醫學健康專家團隊合作，由學習解構營養標籤至隱藏的陷阱，我和各式醬汁轉變成「陌路以上，蜜友未滿」的狀態。「陌路以上」的意思是我不會抗拒它們，因各位廚房掌舵人總有「趕時間」、「想洗少啲、切少啲」、「星期六、日想輕鬆煮」的情況要兼顧，常備幾款現成調味料或醬汁絕對是方便之選；「蜜友未滿」解作君子之交淡如水，剛剛好就好，恰到好處，互相了解，不用深交，以天然食材製作的醬汁為首選。

　　坊間的醬汁做到味道好、賣相好，當中大有機會隱含高熱量、高鈉和高糖的陷阱。

1. 鹽：現成調味料的鹽含量通常較高，過量攝取鹽可能增加患上高血壓、中風及心臟病等疾病的風險。建議選擇低鈉或無鈉調味料，並注意調味料中鈉含量。

2. 糖：現成調味料的糖含量通常較高，尤其是甜味醬油、甜酸醬、沙茶醬及番茄醬等。過量攝取糖可能導致肥胖、糖尿病及心臟病等疾病。建議選擇低糖或無糖調味料，或使用天然甜味食材代替糖。

3. 添加劑：現成調味料可能含有人工添加劑，如防腐劑、色素、香精等，可能對身體健康造成損害。建議選擇不含添加劑的調味料，或選擇添加劑較少的產品。

特別一提常用的番茄醬，現成人工番茄醬通常由非天然番茄醬、糖、醋、調味料和添加劑等混合製成。這些添加劑可能包括防腐劑、色素及香精等，同時含有較高糖和較高鹽成分，長年累月常吃，有機會對身體帶來壞影響。建議可嘗試自製低糖、低鹽、無添加劑的番茄醬，或選擇以天然番茄製造的番茄醬代替。

選購和使用醬汁時，以下幾點需要留意：

1. 選擇低糖、低鈉和低脂的產品，盡量選擇天然食材製成的醬汁。

2. 認識醬汁的成分表，購買前翻到背面詳細地看清成份，避免選擇添加人工色素、香料和防腐劑等添加劑的產品。

3. 適量使用醬汁，避免過度攝取鈉和糖。避免每天每餐過量使用，點綴一下味道，可享用現成醬汁帶來的方便及美味，亦減少對身體的不良影響。

4. 除醬汁外，可多加使用新鮮的香草和香料，如迷迭香、百里香、羅勒、蒜頭、洋蔥和薑等，增加食物的風味，豐富調味的味道及營養價值。

Meal-Preparation

一 週 備 餐 攻 略

　　每個星期，我都會選一日（約 3-4 小時）做 Meal-Preparation（Meal Prep），可能大家看到要花幾個小時會立即卻步，會覺得「咁麻煩，我都係唔做／做唔到㗎啦！」信我信我信我！這一星期內花的幾個小時，是絕對值得！尤其是兼顧上班的煮人們，好好計劃及實行一週備餐，你會發現，每餐打開雪櫃有食材立即可用，不用煩惱每餐煮甚麼、怎樣醃、花多少時間醃等等，也不用苦惱「不如都係出街食算啦！」這是一種多麼美好的感覺！只要大家一步一步跟着參考，相信你們一定可以做得到，並且做得好！

Step 1：好好計劃，選購食材

第一步由購買食材開始，我會劃分為蔬菜及肉類，肉類分為新鮮或急凍肉類。如果一星期只有放假天才能方便買餸備餐的話，請依着以下種類來計劃餐單。

蔬菜

綠葉菜的存放日期較短，每次不宜購買太多，首 2-3 天可吃綠葉菜，隨後吃豆類、瓜類等。除了街市買到的新鮮蔬菜，也可考慮超市的沙律菜。

以下介紹讓綠葉菜耐存的處理方法：

1. 如一次購買數天蔬菜，我會於菜檔挑選較新鮮、乾爽的蔬菜，回家後處理也方便省時。
2. 回家後，我會立即將不安排當天食用的蔬菜鋪平，用廚房紙印乾水分，挑去已發黃或濕掉腐爛的菜葉。
3. 然後用乾淨廚房紙包好，放入乾爽可重用的膠袋或蔬菜用布袋，放於雪櫃備用。

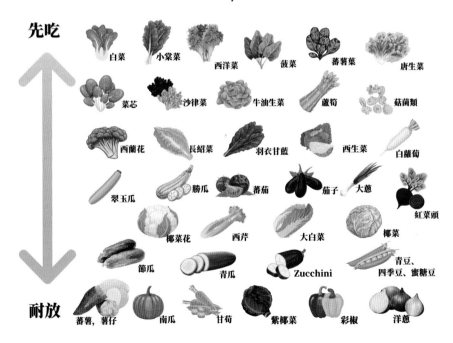

Weekly Plan

先吃

白菜　小棠菜　西洋菜　菠菜　蕃薯葉　唐生菜

菜芯　沙律菜　牛油生菜　蘆筍　菇菌類

西蘭花　長紹菜　羽衣甘藍　西生菜　白蘿蔔

翠玉瓜　勝瓜　蕃茄　茄子　大蔥　紅菜頭

椰菜花　西芹　大白菜　椰菜

節瓜　青瓜　Zucchini　青豆、四季豆、蜜糖豆

耐放

蕃薯, 薯仔　南瓜　甘荀　紫椰菜　彩椒　洋蔥

新鮮肉類及海鮮類

　　於街市肉檔有不少肉類選擇，如新鮮豬扒可選胸頭扒、斧頭扒或水脾，海外讀者可選豬扒芯（Pork centre），或到肉枱選豬扒時挑選連骨部分，並看豬扒有否少許油花，買回來的豬扒只要簡單處理就很好吃了！新鮮排骨可選妃排、一字排、腩排或唐排，我較多選妃排（近肩胛位的排骨），因為帶骨少肥，肉質柔軟，怎樣煮也可新手不敗。

我的 Shopping lists

♥ 豬扒
♥ 胸頭（可做成免治肉，亦可切成肉絲、肉粒等方便炒菜）
♥ 排骨
♥ 新鮮雞髀、全雞
♥ 各種魚
♥ 各種海鮮
♥ 牛肉

冰鮮及急凍肉類

如不講究新鮮肉質及口感，逛凍肉店定能令你滿載而歸，同時舒解心靈的憂慮：「原來減磅真係唔洗捱餓！」只要冰格有足夠的儲存空間就可以了，Let's go！

我的 Shopping Lists

- ♥ **豬扒：**來自世界各地，款式琳琅滿目，每週嘗試一款絕無悶場。
- ♥ **豬柳：**並非快餐店的即食豬柳，而是圓圓長長密封包裝的豬柳，方便做成肉粒、肉絲等。
- ♥ **免治肉：**方便又省時，如有時間我通常用胸頭混合瘦肉打成肉碎。
- ♥ **排骨粒、一字排：**排骨粒可以燜或蒸，西餐常用的一字排（Pork rib）用以烤焗很惹味。
- ♥ **胸肉片、豬腩片：**是打邊爐常用的肉片，非常方便，甚至可免醃肉的步驟，輕鬆煎煮已很好吃。
- ♥ **雞扒、雞柳、雞鎚、雞胸：**煎雞扒；鹽水浸雞柳或雞胸，煮成手撕雞等，無論氣炸、煎、煮、焗都容易處理。
- ♥ **全雞：**有時間的話，可用來蒸雞或焗成脆皮雞等。
- ♥ **牛仔骨、牛扒、牛柳粒、火鍋牛肉片：**除牛仔骨外，其他牛肉相對易於準備，可將牛肉料理放於較忙碌的一天當晚餐。
- ♥ **三文魚、比目魚、銀鱈魚或其他魚柳：**我建議可準備多款密封包裝魚扒，前一晚由冰格取出放於雪櫃解凍，放工回家簡單煎煮就可以。
- ♥ **海鮮：**青口、帶子、蜆肉是我常買的，無論是南瓜海鮮湯或伴味噌湯享用，也非常方便。

大家苦惱怎樣配搭食材，如何計劃一星期甚麼時候吃甚麼，可參考以下的配搭，重點是先吃新鮮魚及肉、醃製肉類可儲存三天左右，週末可考慮吃牛扒、胸肉片、急凍魚扒或海鮮等，我為大家配搭了一週餐單，不妨參考一下：

WEEK	LUNCH	DINNER
星期一 Monday	豬扒	蒸魚 / 白灼蝦
星期二 Tuesday	蒸排骨	煎雞扒
星期三 Wednesday	炒肉絲	蒸肉餅
星期四 Thursday	氣炸牛仔骨	焗雞鎚
星期五 Friday	三文魚	牛扒
星期六 Saturday	南瓜海鮮湯	椰菜肉片鍋

Step 2：備餐換來一週輕鬆煮

我會先計劃每週的菜單，例如想吃香茅豬扒還是蒜蓉豬扒；土魷馬蹄蒸肉餅還是梅菜蒸肉餅等等，然後準備配料，加上必備的薑、葱及蒜等，可開始清洗肉類，用廚房紙吸乾水分，切成喜好的粒狀或條狀。急凍肉類需預先解凍，再用鹽水浸約 30 分鐘以去除雪味。

我的 Meal Prep list 種類很多，有香茅豬扒、蒜蓉或沙薑雞扒、洋葱黑椒豬肋骨、蔬果醃牛仔骨及味噌魚柳等等。

Marinated
Lemongrass Pork Chops
★
香茅豬扒

Ingredients
材料

豬扒 4-5 塊（可選胸頭扒或豬扒芯）
香茅 1-2 枝（視乎粗幼及長短而定）
薑 2 片
葱 1 條
蒜肉 5 瓣
乾葱 3 粒

4-5 pork chops
1-2 lemongrass
(depend on the thickness and the length)
2 slices ginger
1 sprig spring onion
5 cloves garlic
3 shallots

Seasonings
調味料

鹽 2 茶匙
糖少許
料理酒少許
生抽 1 湯匙
麻油 2 茶匙
胡椒粉少許
粟粉 1 1/2 湯匙（後下）
水約 5 湯匙（後下）
油適量（後下）

2 tsp salt
sugar
cooking wine
1 tbsp light soy sauce
2 tsp sesame oil
ground white pepper
1 1/2 tbsp cornstarch (add at last)
5 tbsp water (add at last)
oil (add at last)

Method
做法

1. 豬扒洗淨，用廚房紙吸乾水分，用鬆肉鎚或刀背拍鬆豬扒至約 1.5 倍大（胸頭扒、豬扒芯或斧頭扒可省略此步驟）。
2. 用刀切斷周邊白色筋位，以免煎豬扒時捲起。
3. 香茅洗淨，切去頭末兩端，撕走表面一至兩層較硬部分，用刀背拍扁頭段，較易出味。
4. 薑、蒜肉及乾葱切粒，與香茅同放入攪拌機打碎，放於大碗，加入調味料拌勻（粟粉、油及水除外）。
5. 豬扒逐一放入大碗，用手在豬扒底面細心地抹勻調味料。
6. 用水調勻粟粉，均勻地抹於豬扒底面，最後加上少許油抹勻。
7. 將豬扒放入玻璃盒（確保容器乾爽）可放雪櫃儲存 3-4 小時或過夜。

1. Rinse pork chops and wipe dry with kitchen paper. Tenderize the pork chop with a meat mallet or the back of the knife until it is about 1.5 times larger its size (slightly less if using boneless pork chop or pork loin chop).

2. Cut off the white tendon around the edges of the pork chop, preventing it from curling while pan-frying.

3. Rinse the lemongrass and remove the head and the end sections. Tear off one or two layers of the tougher outer skin and pound lightly the stem part of the lemongrass with the back of the knife to release its flavour.

4. Finely chop the ginger, garlic and shallot. Put them into the blender or food processor together with lemongrass and blend into puree, then place in the big bowl. Add the seasonings (except corn starch, oil and water) and mix well.

5. Put the pork chop one by one. Rub them thoroughly on the both sides with the seasonings.

6. Mix the corn starch with the water. Spread the corn starch slurry evenly over both sides of pork chops. Add the oil evenly at last.

7. Put the pork chops in the storage container (ensure it is dry). Keep in the refrigerator for 3-4 hours or overnight.

Marinated
Garlic Flavour Chicken Fillet
★
蒜蓉雞扒

Ingredients 材料

急凍雞扒 4 塊
蒜頭 1/2 個
薑 2 片
乾葱 1 粒（可省略）

4 frozen chicken thigh cutlet
1/2 cloves garlic
2 slices ginger
1 shallot (optional)

Seasonings 調味料

鹽 2 茶匙
糖 少許
生抽 1 湯匙
胡椒粉 少許
薑黃粉 2 茶匙
料理酒 1 湯匙
粟粉 2 茶匙
油 適量

2 tsp salt
sugar
1 tbsp light soy sauce
ground white pepper
2 tsp turmeric powder
1 tbsp cooking wine
2 tsp corn starch
oil

Method
做法

1. 急凍雞扒預早一晚由冰格取出，放於冷藏格解凍。
2. 大碗內灑入鹽 2-3 湯匙，加入適量水，放入已解凍雞扒浸約 20 分鐘，去除雪味。
3. 雞扒用水略沖洗，用廚房紙印乾，備用。
4. 蒜頭去皮，用刀拍扁，剁成蒜蓉或以小型攪拌機攪成蓉。
5. 大碗內放入蒜蓉及其他調味料（生粉及油除外）拌勻，均勻地塗抹於雞扒。
6. 若選擇攪拌機處理，建議將其他調味料一併放入攪拌，可輕鬆地混合調味料，完成後直接塗抹雞扒上即可。
7. 最後按次序加入生粉及油拌勻，放於儲存容器內，冷藏儲存。

1. Put the frozen chicken thigh cutlet in the lower shelf of refrigerator at the previous night.
2. Add 2-3 tbsp of salt and enough water in the big bowl. Put in the chicken thigh cutlet and soak for 20 minutes to remove the frozen smell.
3. Then rinse the chicken and wipe dry with the kitchen paper.
4. Remove the skin of the garlic and pat lightly. Finely chop the garlic or blend into puree.
5. Put the chopped garlic and the seasoning in the big bowl (except the corn starch and oil). Mix well. Rub the chicken evenly with the seasoning.
6. If using the blender, put all the seasoning in the blender and mix well easily. You can rub the seasoning on the chicken evenly.
7. Lastly, add corn starch and oil and mix well. Put the chicken thigh cutlet in the container and refrigerate.

Marinated

★

洋葱黑椒豬肋骨

Onion and Black Pepper Pork Ribs

Ingredients 材料

急凍豬肋骨（或一字排骨）2 大條
洋蔥 1 個

2 large pieces frozen pork ribs
1 onion

Marinade 醃料

鹽 3 茶匙	3tsp salt
蒜粉 2 茶匙	2 tsp garlic powder
黑椒碎適量	chopped black pepper
紅椒粉 1 茶匙	1 tsp paprika
糖 2 湯匙	2 tbsp sugar
料理酒 3 湯匙	3 tbsp cooking wine
生抽 2 湯匙	2 tbsp light soy sauce
味醂 2 湯匙	2 tbsp mirin

Method 做法

1. 將解凍的豬肋骨浸水約半小時，略沖水，印乾水分。
2. 洋蔥及醃料放入攪拌機打成蓉，倒進食物儲存盒，放入豬肋骨均勻地塗抹，醃 3-4 小時或過夜。

1. Soak the defrosted pork ribs in the water for 30 minutes. Rinse well and wipe dry.
2. Put the onion and the marinade in the blender until puree, then store them in the food container. Add the pork ribs and rub the marinade evenly. Keep them in the refrigerator for 3-4 hours or overnight.

Marinated

★

蔬果醃牛仔骨

Short Beef Ribs in Vegetable and Fruit Marinade

　　以下分享一個「如何不加鬆肉粉也令肉類鬆軟」的備餐方法，這是我的契媽英姐教我的，她經營茶餐廳多年，深得上水及粉嶺街坊的喜愛。我從小到大最愛吃英記豬扒，長大後英姐知道我熱愛做菜，將處理豬扒的美味秘訣分享給我。如果大家有「豬扒醃極唔軟腍、鞋口」的煩惱，不妨試試這個方法，豬肉及牛肉均適用。

Ingredients
材料

急凍牛仔骨 1 磅（約 4-5 塊）
450 g (about 4-5 pieces)frozen short beef ribs

Seasonings
調味料

生抽 1 茶匙
糖適量
麻油 1 茶匙
胡椒粉適量
料理酒 1 湯匙
1 tsp light soy sauce
sugar
1 tsp sesame oil
ground white pepper
1 tbsp cooking wine

Marinade
蔬果汁醃料

西芹 1/2 條
甘筍 1/2 條
奇異果 1/2 個
蘋果或雪梨　1 個
1/2 stem celery
1/2 carrot
1/2 kiwi
1 apple or Chinese pear

做法

1. 將解凍的牛仔骨浸水約半小時，沖水洗淨，用廚房紙印乾水分。
2. 在骨位間切成小件，並於牛仔骨下方輕切一刀以切斷筋位，煎時不會捲起（做法與處理豬扒相同）。
3. 蔬果汁醃料放入攪拌機打好，倒入食物儲存盒，加入調味料拌勻，排入牛仔骨醃味（如當晚食用可浸約30分鐘；厚身豬扒建議浸2小時或過夜，讓蔬果酵素滲透肉質）。

1. Soak the defrosted short beef ribs in the water for 30 minutes. Rinse well and wipe dry.
2. Cut into three pieces of each ribs. Cut the white tendon at the bottom of beef ribs, preventing curly after pan-frying (same as preparing pork chops).
3. Put the marinade in the blender and blend until puree, then store them in the food container. Add the seasonings and mix well. Place the beef ribs for marinating. Marinade them for 30 minutes if serving at night. I suggest marinating the thick beef ribs for 2 hours or overnight, let the vegetable enzyme to make the ribs soft.

TOP 10 健康食材

懂得選擇天然優質的食材,是邁向減醣健康生活的第一步,以下是 TOP 10 健康食材,日常選購時不妨多加留意。

種籽類
豐富食物纖維
加強飽足感

雞蛋
低卡高蛋白
滿滿飽腹感

莓類水果
豐富花青素、抗氧
化力強另有豐富纖
維及維他命

蔬菜類
多元豐富維他命及礦物質

牛油果
含不飽和脂肪酸
屬需要吸收之優質脂肪

新鮮水果配原味乳酪
取代水果味高糖乳酪
較健康
含有益腸道益生菌

每日一份含鈣質奶類
製品或植物奶製品

宜選擇
豐富蛋白質低脂肪肉類

豆類＋果仁類 #理想口痕之選
豐富蛋白質、優質脂肪
注意果仁類含高熱量 不宜進食過量

芝士類
豐富蛋白質、優質脂肪及多種
微量元素,延長飽足感

Chapter 1

Detox
Water

維 他 命 排 毒 水

Recipes

加入了水果、蔬菜、新鮮香草等
豐富食物纖維和抗氧化物的detox drink，
亮眼吸晴，能吸引大家多喝水，
減少攝取其他高甜飲品。

Seek
magic
every
day

Detox Water 的用處

　　提到要達到理想減磅的結果，學懂食物配搭前先要習慣每天喝足夠水。但一提起飲水，大家已經有數不盡的原因訴說，要每天飲足夠水非常有難度。

　　來吧！有心就事成，想健康沒理由每日任由自己喝幾十粒糖進肚子而面不改容，以下一系列 Detox Water，集齊療癒身心（包括眼球）、打卡 -able、有賣相有實力、水有水味等元素！今天開始KO「淨飲水無味無道」的障礙，為大家分享簡單方便、為清水增加吸引度的另類選擇。

　　Detox Water 是指在水中加入水果、蔬菜、新鮮香草等富含食物纖維和抗氧化食材的飲品。請大家不要斷定喝 Detox Water 可以直接幫助減肥；但如果這些「有味水」可以吸引大家多喝水，更甚可以取代對高甜手搖飲料或汽水、果汁的渴求，那麼它絕對可以幫助減少每天的熱量攝取，進而有助體重管理。

　　要研究 Detox Water，細分各食材的功效、營養素於水中釋放的溫度及時間，可以用一本書的時間去記錄，初次為大家介紹，就以日常手到拿來及慣用的幾款食材與大家分享吧！

以下是 Detox Water 的好處：

1. 促進身體排毒：添加檸檬、薑或薄荷等食材，具有利尿和排毒的作用，有助排出身體的毒素和廢物，亦可去除疲勞及提升好心情。
2. 增強代謝力：添加莓類水果、奇異果或蘋果醋等食材，具有促進代謝的作用，有助加速燃燒卡路里和脂肪，更有美肌之效。
3. 增強免疫力：添加莓類水果、薑或肉桂等食材，具有抗炎和抗氧化的作用，有助增強免疫力，保持身體健康。
4. 消除水腫：添加西瓜、蘋果和青瓜等食材，具有利尿和消除水腫的作用，有助減少身體的浮腫感，夏天飲用更可清熱解暑。

Ingredients

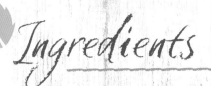

Fruits

＊可隨喜好添加各水果

Herbs

＊隨喜好加添新鮮香草

Honey

＊可選較低升糖指數蜜
　糖，如龍舌蘭蜜、楓
　糖漿
＊蜜糖可省略

Water

＊清水
＊有氣礦泉水

Tea ＊Optional

＊可添加喜好味道的茶包
＊添加熱水

維他命排毒水

49

Option 1

Cucumber and Lemon with Mint

★

Detox Water

青瓜檸檬
薄荷水

Ingredients
材料

青瓜 1/2 條　　　1/2 cucumber
檸檬 1/2 個　　　1/2 lemon
薄荷葉 5 片　　　5 mint leaves
水 1 公升　　　　1 litre water

Method
做法

1. 青瓜及檸檬洗淨，切片。
2. 將青瓜、檸檬及薄荷葉放入容器，注入水，待一會即可飲用。
1. Rinse cucumber and lemon. Cut into slices.
2. Put all the ingredients in the container. Let them soaking for a while and serve.

Skye's tips

添加檸檬或薄荷等食材，具有利尿和排毒
的作用，有助排出身體的毒素。

Option 2

Grapefruit and Cinnamon
★
Detox Water

西柚肉桂
排毒水

Ingredients
材料

西柚 3 片　　　　3 slices grapefruit
肉桂條 1 條　　　1 cinnamon stick
水或有氣水 1 公升　1 litre water or sparkling water

Method
做法

1. 西柚洗淨，切片。
2. 將所有材料放入容器，注入水，待一會即可飲用。
1. Rinse and slice grapefruit.
2. Put all the ingredients in the container. Let them soaking for a while and serve.

Skye's tips

西柚有助排毒消脂，配搭肉桂條減低嘴饞的感覺。

維他命排毒水

Option 3

Green Apple and Thyme
★ Detox Water

青蘋果百里香
排毒水

Ingredients
材料

青蘋果 1 個	1 green apple
或百香果 1 個	or 1 passion fruit
百里香 5 條	5 sticks thyme
水或有氣水 1 公升	1 litre water or sparkling water

Method
做法

1. 青蘋果洗淨，切薄片。
2. 將所有材料放入容器，注入水，待一會即可飲用。
1. Rinse and slice thinly green apple.
2. Put all the ingredients in the container. Let them soaking for a while and serve.

Skye's tips

這個配搭新鮮感滿分，百里香抗氧能力強，泡熱水風味更佳。

Option 4

Watermelon and Mint
★
Detox Water
西瓜薄荷
有氣水

Ingredients
材料

西瓜 6-8 片　　　6-8 slices watermelon
薄荷葉 5 片　　　5 mint leaves
水或有氣水 1 公升　1 litre water or sparkling water

Method
做法

1. 西瓜去皮，切片。
2. 將所有材料放入容器，注入水，待一會即可飲用。
1. Remove the skin of watermelon and cut into pieces.
2. Put all the ingredients in the container. Let them soaking for a while and serve.

Skye's tips

夏日清新之選，當然直接吃西瓜絕對是理想選擇。

維他命排毒水

53

Option 5

Berries and Basil
★
Detox Water

莓果羅勒
排毒水

Ingredients
材料

士多啤梨 4 粒
藍莓 10 數粒
黑莓 5 粒
羅勒 6-8 塊
熱水或水 1 公升

4 strawberries
10 blueberries
5 black berries
6-8 pieces basil leaves
1 litre boiling water or water

Skye's tips

我建議加茶包飲用，用熱水浸泡，味道更
佳。

 *Method*
做法

1. 士多啤梨、藍莓及黑莓用水浸泡，沖淨。
2. 將全部材料放入容器，注入水，待一會即可飲用。

1. Soak all berries in the water and rinse well.
2. Put all the ingredients in the container. Let them soaking for a while and serve.

Option 6

菠蘿
+
青瓜

先將兩者打成汁

菠蘿青瓜汁

檸檬

薑汁

橙

蘋果醋

Pineapple, Cucumber, Orange with Apple Cider Vinegar and Ginger

★ Detox Water

菠蘿青瓜橘橙薑汁蘋果醋水

Ingredients
材料

青瓜 1/2 條	1/2 cucumber
菠蘿 3 塊	3 pieces pineapple
橙 1 個	1 orange
檸檬 1 個	1 lemon
薑汁 2 茶匙	2 tsp ginger juice
蘋果醋 30 毫升	30ml apple cider vinegar
水 1 公升	1 litre water

Method
做法

1. 青瓜洗淨，切片；與菠蘿放入攪拌機打成汁。
2. 橙及檸檬洗淨，切片。
3. 將全部材料放入容器，注入水，待一會即可飲用。

1. Rinse and slice cucumber. Put it into the blender with pineapple.
2. Rinse orange and lemon. Cut into slices.
3. Put all the ingredients in the container. Let them soaking for a while and serve.

Skye's tips

這個絕對是派對之選，用以配搭 party food，非常合適。

維他命排毒水

do something
GOOD TODAY

Chapter 2

Overnight Oats

一 試 愛 上 的 「 隔 夜 燕 麥 」

Recipes

將鋼切燕麥或原片大燕麥泡在奶類製品一晚後，

配上自家喜歡的水果、堅果及種籽等，

飽足感豐富，而且顏色繽紛，

作為營養早餐或午後小吃，

令整天幸福滿滿！

　　隔夜燕麥杯做法簡單，口味多變，零廚藝毋須技巧就可完成，更是 Grab & Go 的最佳選擇。如果家中有小孩，更可邀請他們一同製作，讓他們選擇心儀的口味，以隔夜燕麥杯當他們的早餐、課後下午茶或零食甜點都不錯啊！

　　說得這麼簡單，究竟怎樣做呢？

1. 睡前把燕麥和其他喜歡的食材放進玻璃樽攪拌。
2. 放入雪櫃。
3. 翌日起床加入新鮮水果，立即有美味又飽肚的早餐吃。
4. 就是這麼簡單啊！是不是很驚喜，很想知道放甚麼材料最好吃呢？來吧！花幾分鐘時間看看隔夜燕麥是甚麼吧！

選哪款燕麥適合？

　　燕麥片不只有一種，仔細看英文名稱，你會發現燕麥除了有不同的名稱、不同的呈現狀態，最為大家熟悉的是用來沖或煮「麥皮」早餐的快熟燕麥（Quick Oats）及即食燕麥（Instant Oats）兩種，還有即沖燕麥飲品。這幾類燕麥相對加工較多，升糖值較高，同時注意即沖燕麥飲品大部分已被預先調味或有添加糖。

　　我建議大家選擇原片大燕麥（Rolled Oats），甚至找到鋼切燕麥（Steel Cut Oats）更為理想。這兩類屬較少加工的燕麥，營養價值如纖維會更完整地保留，是製作隔夜燕麥的理想選擇。

甚麼是隔夜燕麥？

　　「隔夜燕麥」就是將燕麥泡在奶類製品（或隨口味選擇喜好的液體），浸泡一晚後，燕麥被軟化成充滿咀嚼感，同時吸滿香濃奶香近似軟糯布甸的口感。浸泡「隔夜燕麥」最常用的是全脂牛奶、低脂牛奶、脫脂牛奶、豆奶、各類植物奶或原味無添加糖乳酪等，再加入各種配料如奇亞籽、水果、堅果、種籽類、乾果都可以，放入雪櫃冷藏一晚，翌日只需隨習慣或喜好添加新鮮水果就可以了，或可成為帶到辦公室的早餐或午餐之選，更可成為運動後的補給小食。

隔夜燕麥可搭配以下早餐：

1. 焓蛋、煎蛋、炒蛋——補充足夠蛋白質，增加持續飽足感。
2. 牛油果——優質油脂、豐富膳食纖維、低升糖水果，豐富飽足感。
3. 飲品可選配無添加糖黑咖啡、無糖檸檬茶或綠茶。

隔夜燕麥的 3 大好處

1. **增加飽足感**：水溶性膳食纖維能增加飽足感，可避免正餐進食過量，減少餐與餐之間的零食口癮，有助控制總熱量的吸收。
2. **改善腸道健康**：燕麥含豐富水溶性膳食纖維，促進腸胃蠕動，加上飲用足夠水分，有潤腸通便、排出毒素及廢物的作用。
3. **豐富營養**：燕麥含多種營養，包括蛋白質、纖維、多種維他命及礦物質，以及人體所需的氨基酸。

如何準備隔夜燕麥？

1. 儲存容器：任何玻璃樽或杯

建議大家使用玻璃容器，可選一個適合自己食量的杯，在家我會常備不同大小的梅森罐（Mason Jar），我喜歡它的外形，也方便我加添不同食材，視覺上更吸睛，方便打卡（笑）。

2. 基底食材：

燕麥、奇亞籽、奶類製品（奶、乳酪或植物奶）、蜜糖或楓糖漿（隨意）。

3. 可配食材：

時令水果、堅果、種籽、乾果、自家煮低糖果醬。

4. 額外加配：

亞麻籽粉、黑朱古力粉、黑芝麻粉、黑豆粉、肉桂粉、椰子碎、燕麥糠、膠原蛋白粉、蛋白粉（如欲增加或補給蛋白質）。

好處：

1. 純天然糖分，含糖量較低約 60%（蜂蜜約為 80%）

2. 升糖值相對其他蜜糖及糖為低

3. 熱量相對其他蜜糖及糖為低

4. 含有豐富礦物質

5. 含有豐富抗氧化物

6. 提升人體免疫力

Basic combo

Rolled Oats or
Steel Cut Oats

Chia Seeds

Milk of Your Choice
Dairy or Plant Base

Honey or Maple Syrup

Base

Base of Overnight Oats

★

隔夜燕麥基底

Ingredients

材料

原片大燕麥 4 湯匙
奇亞籽 2 茶匙
牛奶 200 毫升
蜜糖或龍舌蘭糖漿 少許

4 tbsp rolled oats
2 tsp chia seeds
200 ml milk
honey or agave nectar

＊工具 Utensil
梅森罐 1 個（300-500 毫升）
1 Mason jar (300-500ml)

Method
做法

1. 將燕麥、奇亞籽放進玻璃樽，加入牛奶攪勻（如想燕麥口感較厚實，可添加少許乳酪，或翌日進食前加入），放入雪櫃冷藏一晚，成為隔夜燕麥基底。

2. 蜜糖或龍舌蘭糖漿可隨口味加添；如加入香蕉或芒果等相對甜度較高的水果，盡量避免添加蜜糖。

3. 隔夜燕麥基底冷藏一夜後，配上喜愛的水果、果仁或種籽等，一齊來享受健康的早餐。

1. Put rolled oats and hemp seeds into the mug. Add the milk and stir well (add some yoghurt if you want to taste thick oats, or you can add yoghurt when serving next day). Keep it in the refrigerator overnight. It is the base of overnight oats.

2. According to your taste, add honey or agave nectar. When you mix with sweet fruit like banana or mango, suggest that it is not to add honey.

3. After refrigerated the overnight oats, add the fruit, nuts or seeds as you like next morning. It's the healthy breakfast for you!

Skye's tips

燕麥基底可放於雪櫃三至四天，進食前取出添加水果、乳酪等就可以了。

奇異果・士多啤梨・大麻籽・開心果
kiwi・strawberry・hemp seeds・pistachio

★

基礎版 2

黑莓・芒果・合桃・葵花籽
blackberry・mango・walnut・sunflower seeds

一試愛上的「隔夜燕麥」

Basic 5

★ 基礎版 3

香蕉・藍莓・杏仁・松子仁
banana・blueberry・almond・pine nuts

Base of Overnight Oats Variation

★ 隔夜燕麥 變奏版基底

*Method*
做法

1. 將香蕉、黑朱古力粉 / 紅肉火龍果 / 芒果放進玻璃樽壓成蓉，加入隔夜燕麥基底攪勻，放入雪櫃冷藏一晚，成為不同的變奏版基底。
2. 放上乳酪及自己喜愛的水果、果仁、種籽及乾果。

1. Put banana and cacao powder/red dragon fruit/mango in the mug and press into puree. Add the overnight oats base and stir well. Keep in the refrigerator overnight. It is the base of overnight oats variation.
2. Top with yoghurt, fruits, nuts, seeds and dried fruit as you like.

Skye's tips

> 習慣弄基礎版的隔夜燕麥，就可以嘗試加不同的水果及味道。

Ingredients
材料

水果（香蕉 1/2 條、黑朱古力粉 1 湯匙 / 紅肉火龍果 1/2 個 / 芒果 1/2 個）
隔夜燕麥基底
乳酪適量
任何喜好的果仁、種籽、乾果

fruits (1/2 banana and 1 tbsp cacao powder/1/2 red dragon fruit/1/2 mango)
overnight oats base
yogurt
nuts, seeds and dried fruit

一試愛上的「隔夜燕麥」

69

Variation 1

★
變奏版 1

芒果・杏仁・椰絲・乳酪
mango・almonds・
desiccated coconut・yoghurt

Variation 3

★
變奏版 3

香蕉・葵花籽・合桃・乳酪
banana・sunflower seeds・
walnut・yoghurt

Variation 2

★
變奏版 2

紅肉火龍果・松子仁・乳酪
red dragon fruit・
pine nuts・yoghurt

Chapter 3

Tsukemono

漬　物
Recipes

家裏雪櫃常備多款膳食纖維豐富，
兼有營養的低鈉低糖版漬物，
隨時用於便當或減醣餐膳，
成為綠色蔬菜以外的彩虹之選。

「漬物」一詞，可能對於不太熟悉日、韓、台料理的你來說，有機會感到陌生。明明是港式料理，漬物並不普及，何解放它們遠遠於其他餸菜之前呢？說到要為大家分享漬物的緣起，來源於以下幾點：

1. 漬物實在是非常簡單、健康美味，同時易於準備、可常備雪櫃的菜式，它絕對值得擁有更多關注度！
2. 漬物的蔬菜選擇種類眾多，多備幾款，手到拿來可以解決大家經常認為除了綠色蔬菜，不知還有甚麼好吃的疑慮。
3. 這些當然是我自己十分喜愛漬物，實在希望大家可以參考幾款我最經常準備的配搭，隨手輕鬆做出營養、膳食纖維豐富的低鹽低糖版漬物！

以下各款漬物毋須用大量鹽和糖醃漬，食譜口味屬偏清淡，建議大家自行調節醬汁分量，以及可添加喜好之蜜糖、龍舌蘭糖漿或楓糖漿等。

另一溫馨提示：處理漬物前，記得容器要做足準備功夫，確保漬物放於已消毒乾淨及烘乾的密封玻璃瓶、或食物儲存容器，以得到最佳保存效果；取出漬物時使用潔淨及乾爽的器具，不宜選擇已使用於進食或非潔淨的餐具。

料理界寶物——鹽麴、醬油麴

　　這個料理界的魔法寶物一定要向大家分享！

　　鹽麴（日文為塩麴、塩糀 Shiokouji，しおこうじ），最早出現於日本江戶時代，江戶時代文獻《本朝食鑑》曾有記載，是一種經過發酵的調味料。到了平成時代（1989-2019）由大分縣一家麴酒釀造廠開始推出家用的鹽麴，從而推及至全國。製造鹽麴的材料簡單及天然，由米麴、鹽和水經過發酵而成，毋須加添其他添加物，而且製造過程簡單，在家輕鬆按步驟就可完成。

　　鹽麴是發酵食物，它擁有一百多種活性酵素，煮食時可取代鹽，食物的成品比加鹽沒有硬繃的鹹味，經發酵後的鹽麴可以帶出鹽的鹹味、（來自澱粉）醣類的淡淡甘甜味及散發發酵食物特有的甘醇深厚，完全可以提升食物的美味層次，為你的愛心料理增添滋味。

鹽麴的好處

1. 鹽麴是發酵食物，麴菌生長時會產生發酵代謝物等，有利於皮膚、腸道健康及有助調節新陳代謝。
2. 鹽麴產生豐富的酵素可將蛋白質分解成氨基酸，醃肉時就像 24 小時有以萬計技師為肉類按摩，令肉質香滑鬆軟，而且用鹽麴醃肉可延長肉質保存期。
3. 鹽麴的鈉含量比食鹽低，可減少腎臟負擔，同時毋須捨棄食物的鮮味享受，是代替食鹽的一個好選擇。
4. 鹽麴含有澱粉酶、蛋白酶及脂肪酶等消化酵素，幫助消化。

鹽麴用法

1. 醃肉提鮮。
2. 醃魚去除腥味。
3. 取代鹽炒菜、煮湯等。
4. 涼拌菜或製作漬物。
5. 基本上各款用鹽調味的料理可用鹽麴替代。

Shiokouji
The Making of Shiokouji

★
鹽麴
製作方法

Ingredients
材料

米麴 300 克   300 g koji rice
海鹽 100 克   100 g sea salt
常溫飲用水 500 毫升   500ml drinking water
(at room temperature)

Method
做法

1. 準備一個密封玻璃瓶（因發酵時產生氣泡，建議選一個所需容量較高身的玻璃瓶），消毒乾淨，烘乾備用。
2. 按照包裝指示，將米麴及海鹽放入已消毒烘乾的玻璃瓶。
3. 加入飲用水，用乾淨的湯匙或筷子拌勻（防止可能引致黴菌及細菌），攪拌後輕輕蓋上蓋，讓米麴有空間呼吸，冷藏。
4. 隨後連續 7-10 天，每天打開蓋，以乾淨湯匙攪拌，直至糊化成稀粥狀（如潮州粥模樣）就可使用。

1. Prepare a sealed bottle (suggest to choose the higher capacity glass bottle). Sterilize the bottle and make sure it is dry enough.
2. Follow the instruction, put the koji rice and sea salt in the glass bottle.
3. Add the drinking water and stir with the clean spoon or chopsticks (to avoid the bacteria). Cover the lid and let the koji rice for fermentation. Store in the refrigerator.
4. Open the lid and stir well with the clean spoon every day within 7-10 days, until it is thicken. It is ready for seasoning.

Shiokouji

The Making of Shiokouji with Soy Sauce

★
醬油麴
製作方法

Ingredients
材料

米麴 300 克　　　300 g koji rice
醬油 600 毫升　　600ml soy sauce

Method
做法

做法與鹽麴做法相同。
The method is same as
making Shiokouji.

Skye's tips

大家毋須兩款同時做，可先做鹽麴試試配
搭菜式，習慣使用後再試做醬油麴。

Lotus Root with Yuzu Honey Sauce

柚子蜜汁
涼拌蓮藕

Ingredients
材料

蓮藕 約 1/2 條	1/2 lotus root
柚子蜜 2-3 湯匙	2-3 tbsp yuzu honey
檸檬皮 少許	lemon zest
迷迭香適量	rosemary
暖水適量	warm water

Method
做法

1. 蓮藕洗淨，去皮、切片。
2. 鍋內加入水，煮滾後加少量鹽，放入蓮藕以小火煮 5 分鐘，盛起，放入
   冰水中。
3. 於常備菜盒內，加入柚子蜜以暖水拌好，刨少許檸檬皮，加入迷迭香拌
   勻，再加入蓮藕片，確保每片蓮藕片均勻地浸於柚子蜜汁。
4. 將常備菜盒蓋好放於雪櫃，翌日即可享用。

1. Rinse and peel the lotus root, then slice it into rounds.
2. Bring a pot of water to a boil, add a pinch of salt. Cook the lotus root over
   low heat for 5 minutes. Set aside and cool them down in ice water.
3. In a container, mix together the yuzu honey with warm water. Add lemon zest,
   rosemary and the lotus root slices in the container, make sure each slice of
   lotus root is evenly coated with the sauce.
4. Cover the container and place it in the refrigerator. Enjoy the dish the next day.

Skye's tips

相中的紫色水來自紫椰菜，若碰巧煮紫椰
菜，留起紫椰菜水浸泡蓮藕，可成為漂亮
的紫色蓮藕片。

Japanese-style Pickled Purple Cabbage

和風
紫椰菜漬

Ingredients
材料

紫椰菜 1/4 個	1/4 purple cabbage
甘筍 1/2 條	1/2 carrot
木魚碎小半碗	a small bowl bonito flakes
白芝麻適量	white sesame seeds
百里香適量	thyme

Seasonings
調味料

米醋 100 毫升（視乎椰菜分量）
楓糖漿或龍舌蘭糖漿 約 30 克
橄欖油 2 湯匙
芥末籽醬 2 茶匙
100ml rice vinegar
(depend on the quantity of cabbage)
30 g maple syrup or agave nectar
2 tbsp olive oil
2 tsp whole grain mustard

Method
做法

1. 紫椰菜、甘筍洗淨，切絲。
2. 大碗內加入食鹽 1/4 茶匙，放入紫椰菜絲輕輕抓拌，靜置 10 分鐘去青味。
3. 如不放心生食者，可將紫椰菜絲放入滾水快灼 30 秒，瀝乾備用。
4. 倒掉去青水後，放入冰水浸約 15 分鐘，可提升椰菜絲的清脆度，瀝乾水分備用（怕麻煩者可省略此步驟）。
5. 在常備菜盒內放入所有調味料，調味後，加入紫椰菜絲及甘筍絲拌勻即可，進食前灑上木魚碎或白芝麻（可略）。

1. Rinse the purple cabbage and carrot. Cut them into thin strips.
2. In a large bowl, mix the purple cabbage with 1/4 tsp of salt and let it sit for 10 minutes to remove the bitterness.
3. If you are concerned about raw consumption, blanch the vegetables in boiling water for 30 seconds, then drain and set aside.
4. Drain the water and soak the vegetables in ice water for about 15 minutes to improve the crunchiness. Drain the excess water. This step is optional.
5. In a bowl or storage container, mix all the seasoning ingredients together. Add the purple cabbage and carrot strips and mix well. Sprinkle bonito flakes or white sesame seeds on top before serving (optional).

漬物

Refreshing Spicy Cucumber

消暑
微辣青瓜

Ingredients
材料

青瓜 2 條　　　2 cucumbers
紅辣椒適量　　　red chilli
蒜頭適量　　　　garlic

Seasonings
調味料

原糖 / 蜜糖 30 克　　30 g raw sugar/honey
（口味自行加減）　　(the taste can be
米醋 100 毫升　　　　 adjust by yourself)
麻油及鹽各少許　　　100ml rice vinegar
　　　　　　　　　　sesame oil
　　　　　　　　　　salt

Method
做法

1. 青瓜斜切，然後上下反轉再斜切（兩次不要一刀切到底），平放青瓜切成小段。
2. 青瓜放入保鮮袋，灑入鹽 1 湯匙，輕力搓勻，靜待 10 分鐘後，濾出多餘水分。
3. 蒜頭拍扁，切成蓉；紅辣椒去籽、切碎，放於碗內，拌入米醋及糖。
4. 在碗或常備菜盒放入青瓜，加入拌勻的調味料及麻油即可。

1. Cut the cucumber diagonally on both sides without cutting all the way through. Cut the diagonally sliced cucumber into small pieces.
2. Put the cucumber in a plastic bag with 1 tbsp of salt, gently rub to distribute the salt evenly. Let the cucumber sit for 10 minutes, then drain off any excess water.
3. Pat the garlic and chopped. Remove the seeds of red chilli and finely chopped. In a bowl, mix the chopped garlic, chopped red chilli, rice vinegar and raw sugar together.
4. Add the cucumber and sesame oil to the bowl and mix well.

Skye's tips

糖及醋的比例可按口味而加減，一般可用醋 2-3 份及糖 1 份作為參考。

漬物

Onion in Vinegar

醋漬洋葱

Ingredients
材料

純釀米醋、意大利黑醋
洋葱、紫洋葱各 1 個
楓糖漿
鹽

pure rice vinegar

balsamic vinegar

1 onion

1 purple onion

maple syrup

salt

Method
做法

1. 洋葱去外衣,切薄片。
2. 儲存洋葱的容器以高溫消毒,抹淨。
3. 於保存容器中倒入純釀米醋、蜜糖和鹽,攪拌至鹽及蜜糖溶解,以無沾水的筷子夾入紫洋葱,均勻地浸於醋。
4. 洋葱及黑醋存放於另一個容器,拌勻。
5. 加蓋密封,冷藏 5 天即可食用。

1. Remove the outer skin of the onion, slice thinly.
2. Clean and sterilize the storage container with high temperature before use.
3. Pour rice vinegar, maple syrup and salt into the storage container. Stir until the salt and maple syrup dissolve completely. Use chopsticks that are free of water to pick up the purple onion slices and soak them evenly in the vinegar mixture.
4. Place onion slices and balsamic vinegar in the other container, then mix well.
5. Seal the container with a lid and refrigerate for 5 days before consuming.

Skye's tips

醋及糖的分量視乎洋葱分量及容器大小而加減,一般糖及醋的比例 1 比 1 來調味,亦可加香草來提味。

意式
油漬彩椒

Italian-style Oil-pickled Bell Peppers

Ingredients
材料

三色甜椒各 1 個	1 bell pepper of each colour (in different colour)
橄欖油適量	olive oil
迷迭香適量	rosemary
鹽及黑椒碎適量	salt and chopped black pepper

Method
做法

1. 將甜椒放於爐火上烤，每隔一段時間翻面，讓甜椒各面平均受熱。
2. 將甜椒放入保鮮袋，或放入碗並蓋上保鮮紙，讓甜椒的餘溫散出蒸氣軟化外皮，比較容易撕除。
3. 將外皮及籽摘除，較難清除的黑渣輕輕地用水沖淨，為了保留香氣，盡量避免持續水洗。清掉彩椒的黑色外皮，彩椒肉切成條狀。
4. 在碗或常備菜盒內加入甜椒、鹽、黑椒碎、香草，倒入橄欖油浸過所有食材，加蓋，放入雪櫃冷藏，隔天可食用。

1. Grill the bell peppers on a gas stove. Be sure to turn them over every once in a while to ensure even heating.
2. Put the bell peppers in a plastic bag or a bowl covered with plastic wrap to let the steam soften the skin. This makes it easier to peel the skin off.
3. Carefully remove the skin and seeds. Rinse off any black residue that is difficult to remove with water. To preserve the flavour, avoid washing the bell peppers for too long.
4. In a bowl or container, mix the bell peppers with salt, black pepper, rosemary and basil, cover all ingredients with olive oil. Cover the container and refrigerate overnight before consuming.

Skye's tips

油及調味料的分量沒有固定，請按容器大小及食材分量自行調節。

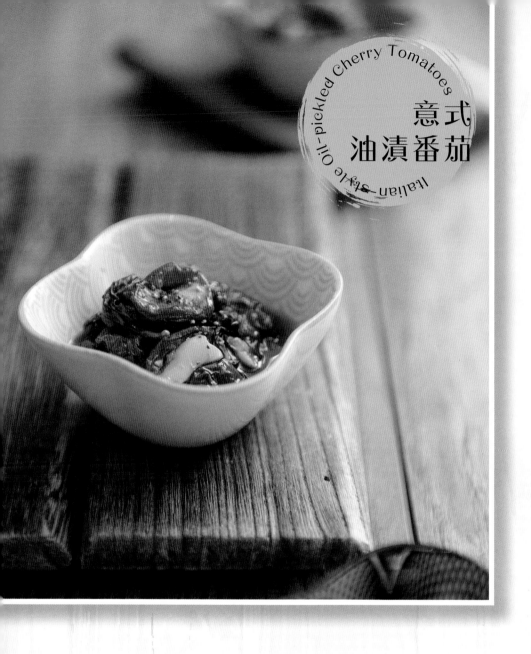

意式
油漬番茄

Italian-Style Oil-pickled Cherry Tomatoes

Ingredients
材料

車厘茄 2 碗（可混合不同顏色）	2 bowls cherry tomatoes
蒜頭 1/2 個	(mix different colours)
乾辣椒或泡辣椒適量	1/2 garlic cloves
九層塔適量	dried red chilli or fresh red chilli
檸檬汁適量	basil
	lemon juice

Seasonings
調味料

橄欖油	olive oil
黑椒碎	chopped black pepper
鹽	salt

Method
做法

1. 蒜頭去皮，拍扁；車厘茄洗淨，切半，同放入焗盤，灑上適量橄欖油、黑胡椒及鹽拌勻。

2. 焗爐預熱 130℃，放入車厘茄焗 50 分鐘或至水分烤乾，水分愈少可保存愈久，待涼。

3. 烤乾車厘茄與蒜片放於容器內，加入新鮮九層塔及乾辣椒，或隨個人口味加添其他香草，灑上檸檬汁。

4. 倒入適量橄欖油浸過所有食材，置於雪櫃冷藏保存，翌日食用風味更佳。

1. Peel and crush the garlic cloves. Rinse the cherry tomatoes and cut them into halves. Place them in a baking dish. Drizzle a moderate amount of olive oil over the tomatoes, then add black pepper and salt. Mix well.

2. Preheat the oven to 130°C and bake the cherry tomatoes for 50 minutes or until they are dried out. The less moisture there is, the longer they can be stored. Let them cool down to room temperature.

3. Place the dried cherry tomatoes and garlic in a container. Add fresh basil and dried red chilli or any other herbs to taste. Sprinkle with lemon juice.

4. Pour enough olive oil to cover all the ingredients, refrigerate overnight for best flavour.

Skye's tips

除橄欖油外，可用牛油果油作為浸泡油，也可轉用其他香味食油來換換口味。

和風淮山
伴秋葵

Japanese Yam and Okra with Japanese-style Dressing

Ingredients
材料

日本淮山 1/3 條	1/3 Japanese yam
秋葵 5-6 條	5-6 okras
大葉（紫蘇葉）5 片	5 Shiso leaves
蝦皮 1/2 碗	1/2 bowl dried tiny shrimp
紫菜碎 適量	dried seaweed flakes
青檸汁 2 茶匙	2 tsp lime juice
木魚碎適量（可省略）	bonito flakes (optional)
黑芝麻適量（可省略）	black sesame seeds (optional)

Seasonings
調味料

紫蘇醋	Shiso vinegar
龍舌蘭糖漿或	agave syrup or
楓糖漿（可省略）	maple syrup (optional)
牛油果油 2 湯匙	2 tbsp avocado oil

Method
做法

1. 秋葵去蒂，用鹽輕力搓洗，放入滾水快灼 1 分鐘，盛起，浸泡冰水降溫。
2. 淮山洗淨，去皮、切塊（緊記配戴手套避免皮膚痕癢）。
3. 大葉切碎；蝦皮用白鑊炒香，盛起備用。
4. 在碗或常備菜盒內放入食材，加入紫蘇醋及檸檬汁及牛油果油拌勻。
5. 進食前灑上木魚碎或黑芝麻；如喜好可略添加龍舌蘭糖漿或楓糖漿以增添風味。

1. Remove the stems from the okra and rinse them with salt. Quickly blanch the okra in boiling water for 1 minute, then cool down in ice water .
2. Peel and cut the Japanese yam into chunks (remember to wear gloves avoiding itching skin).
3. Finely chop the Shiso leaves and set aside. Toast the dried tiny shrimp in a dry pan until fragrant, set aside.
4. In a bowl or container, mix together the ingredients, add shiso vinegar, lime juice and avocado oil..
5. Before serving, sprinkle bonito flakes or black sesame seeds on top. For added flavour, you can also add agave syrup or maple syrup.

Skye's tips

除了紫蘇醋外，也可用柚子醋，兩者於日式超市有售。

漬物

醬油麴漬蘿蔔

Pickled Radish with Soy Sauce Koji

Ingredients

材料

白蘿蔔 1/2 條	1/2 radish
薑適量	ginger
乾昆布 （手掌大小）	dried kelp (the size of your palm)
鹽適量	salt

Seasonings
調味料

醬油麴（做法見 P75）	soy sauce koji (refer to P75
檸檬皮少許	for the method)
檸檬汁適量	lemon zest
鹽	lemon juice
原糖或蜜糖	salt
米醋	raw sugar or honey
	rice vinegar

Method
做法

1. 白蘿蔔洗擦乾淨，去掉外皮，切小薄塊或小粒。
2. 大碗內放入蘿蔔，加入鹽醃漬蘿蔔 10 分鐘，擠出水分。
3. 調味料煮沸，放涼備用。
4. 薑去皮，切細絲或磨成薑汁；乾昆布用廚房紙輕擦，剪成幼絲備用。。
5. 白蘿蔔連同薑絲或薑汁、乾昆布放於漬物容器內，倒入已放涼的調味料，放雪櫃待一晚即可食用。

1. Remove radish skin and cut into thin slices or small cubes.
2. Place the radish in a large bowl, add salt and let it sit for 10 minutes to draw out excess water.
3. Boil the seasonings in a pot, then let it cool.
4. Peel and slice the ginger into shreds or squeeze ginger juice. Wipe the dried kelp and cut into slices..
5. Put the sliced radish and ginger shreds or ginger juice, along with the dried kelp, into a pickling container. Pour in the cooled seasoning mixture. Refrigerate overnight before serving.

Skye's tips

各漬物的糖及醋分量，可自行調整。

漬物

希望大家都找到
自己喜歡的漬物口味！

Chapter 4

Low-Carb Bento

日 常 減 醣 便 當

Recipes

減醣食材絕對適合做便當，

記住以下 3 個重點：

1. 足夠蛋白質

2. 多樣化蔬菜

3. 優質澱粉質

加上做好備餐功夫，可輕鬆做出豐富的便當。

認識便當新拍檔——冷凍蔬菜

　　想於忙碌生活、緊密的工作日程、繁重家務中，同時兼顧準備自家製午餐盒，你要認識一個好幫手——冷凍蔬菜。想起家中長輩常説：「新鮮蔬菜隨手買到，當然要吃新鮮的！」對！放工後有足夠時間買菜的話，選新鮮的、吃新鮮的固然理想，若然當天買不到或沒有時間購買，冷凍蔬菜可説是一個「稱職的」選擇。

　　冷凍蔬菜的選擇有很多，作為餐盒配菜非常不錯，準備時間短，簡單方便，你應該再無任何藉口不吃蔬菜了吧！可參考以下例子：

1. 秋葵
2. 枝豆
3. 西蘭花
4. 椰菜花
5. 粟米粒
6. 青豆粒
7. 抱子甘藍（椰菜仔，Russel sprouts）

玉子燒變奏版

　　玉子燒絕對是午餐便當的神隊友，除了提供滿滿的蛋白質外，還有幸福滿滿的視覺觀感（倩揚視角 ^^）。以下為大家快閃分享幾款玉子燒的堂兄弟姊妹，大家嘗過原味後，跟着以下變奏版作參考，轉轉口味，你的便當將會更加精彩！

變奏版 1：雞蛋配蝦皮（或櫻花蝦）、葱粒。
變奏版 2：雞蛋配紫菜碎、肉鬆。
變奏版 3：雞蛋配紫菜、芝士。
變奏版 4：雞蛋配甘筍絲、椰菜絲。

WEEKLY MEAL PREP

Lunch Box 澱粉類食物選擇

 紅米　 糙米　 黑米　 紫米

 扁豆　 薏米　 小米　 原片大麥片

 薯仔　 蕃薯　 保溫壺　 電飯煲

 南瓜　 粟米　 微波爐　 焗爐

料理方法：

- 粗糧米類別洗淨後浸3至4小時，放入電飯煲煮熟
- 或可直接按電飯煲雜穀米模式（如有）
- 薯仔、南瓜、蕃薯等可用蒸、焗、燴或烤模式烹調
- 可選擇於前一晚在家處理好，或如辦公室有烹調用具即可於午飯前準備

WEEKLY MEAL PREP

Lunch Box 首選新鮮蔬菜選擇

 椰菜　 西蘭花　 椰菜花　 娃娃菜　 青瓜　 秋葵

 法邊豆　 四季豆　 蘆筍　 大蔥　 翠玉瓜　 包裝菇類

LunchBox 次選急凍或罐頭蔬菜選擇

 沙律菜　 沙律菠菜　 椰菜仔　 豆角　 紅菜頭　 紅腰豆

 牛油生菜　 羽衣甘藍　 枝豆　 青豆　 雜豆　鷹咀豆/黑豆

日常減醣便當

午 餐 便 當 1 ＊一人分量

蛋白質：香茅豬扒

蔬菜：蒜蓉炒翠玉瓜、三色豆

自選常備菜：醋漬洋葱（p.82）、醬油麴漬蘿蔔（p.92）

優質澱粉質：櫻花蝦紅米飯、烘南瓜片

Skye's tips

常備菜是便當的好拍檔，為大家解決了不知怎樣帶蔬菜的煩惱，每天一點點分量，簡便、健康、美味！

Bento 1

Fried pork Chops with Lemongrass

★

香茅豬扒

Ingredients
材料

香茅豬扒（醃製方法參考 p.37）

lemongrass pork chops
(refer to p.37 for prepartion)

Method
做法

1. 煮豬扒前，去掉豬扒表面過多的香茅醃料，以免表面過早煎成焦黑色。
2. 燒熱油鑊，放入豬扒以中大火煎封兩面，以免肉汁流失，加蓋，以小火續煎 10-12 分鐘，期間翻轉一次，上碟。
3. 或選用氣炸鍋，使豬扒上色更均勻。預熱 160℃ 氣炸 12-15 分鐘，期間翻轉，再調高溫度至 180℃ 氣炸 3-5 分鐘或至上色即可。

1. Before cooking the pork chop, it is recommended to remove excess lemongrass from the surface to prevent it from overcook.
2. If you prefer to pan-fry the pork chop, fry both sides of the pork chop over high heat then reduce to low heat and cover the pan. Continue cooking for 10-12 minutes, you can flip once or more until it is cooked.
3. You can choose to air-fry the pork chop. Preheat the air-fryer to 160℃ , cook for 12-15 minutes, flip once. Then adjust the temperature to 180℃ , air-fry for an additional 3-5 minutes or until golden brown. Adjust the cooking time according to the size and thickness of the pork chop.

Bento 1

Stir-fried zucchini with Grated Garlic

★

蒜蓉炒翠玉瓜

Ingredients
材料

翠玉瓜 1 個	1 zucchini
蒜頭 3 瓣	3 cloves garlic
三色豆 1/3 碗	1/3 bowl frozen assorted vegetables
鹽及糖各少許	(green peas, diced carrot and
味醂少許（可省略）	corn kernels)
麻油少許	salt
	sugar
	mirin (optional)
	sesame oil

Method
做法

1. 翠玉瓜洗淨外皮，切片或切粒備用。
2. 蒜頭去皮、拍扁，切粒；三色豆解凍，備用。
3. 油熱起鑊，放入蒜粒炒香，下翠玉瓜拌炒數分鐘後，加入鹽及糖或味醂拌勻，加入三色豆炒勻，熄火，下麻油拌勻即成。

1. Clean the zucchini and then slice or dice.
2. Remove the skin of the garlic. Pat lightly and diced. Defrost the frozen assorted vegetables.
3. Stir-fry the garlic until fragrant. Add zucchini and stir-fry for a few minutes. Sprinkle with salt and sugar (or mirin) for stir-frying. Put in the frozen assorted vegetables and mix well. Turn off the heat and add the sesame oil.

Dried Sakura Shrimps Red Rice with Baked Pumpkin

★

櫻花蝦紅米飯烘南瓜

* 米類一次可多煮一點，
方便準備下一餐。

**Cook more rice at one time
keeping for next meal.**

Ingredients
材料

南瓜 5-6 片	5-6 slices pumpkin
櫻花蝦 1/4 碗	1/4 bowl dried sakura shrimps
紅米 1/2 杯	1/2 cup red rice
糙米 1/2 杯	1/2 cup brown rice
小米、藜麥共 1/2 杯	1/2 cup millet and quinoa

Method
做法

1. 南瓜去皮，切片備用。
2. 糙米及紅米預先浸 3-4 小時。
3. 櫻花蝦、小米、藜麥、糙米、紅米加水同煮，放入電飯煲按煮飯模式。
4. 煮飯模式尚餘 8 分鐘時，放入南瓜片同時蒸煮，或將南瓜片放於焗盤，灑入橄欖油及少許鹽拌勻，以 180℃ 焗 8 分鐘，翻轉再焗 3 分鐘（時間視乎南瓜厚薄而增減）。

1. Peel the skin of the pumpkin and slice.
2. Soak the brown rice and red rice in the water for 3-4 hours in advance.
3. Put the dried sakura shrimps, millet, quinoa, brown rice, red rice and water in the inner pot of rice cooker. Cook the rice to set the regular rice mode.
4. Put in the sliced pumpkin or place the pumpkin in the baking tray with olive oil and a pinch of salt and mix well. Grill for 8 minutes at 180°C, then flip the pumpkin and grill for 3 minutes (adjust the time depending on the thickness of the pumpkin).

午 餐 便 當 2 *一人分量

蛋白質：雞柳
蔬菜：甘筍、木耳、青椒、洋葱
自選常備菜：柚子蜜汁涼拌蓮藕（p.76）、消暑微辣青瓜（p.80）
優質澱粉質：焗薯角

Skye's tips

> 黑色食物大家一般比較少放於日常餐單，
> 木耳是一個非常好的選擇。

102

Bento 2

Baked Potato Wedges

★

焗薯角

Ingredients
材料

薯仔	potato
鹽及橄欖油各適量	salt
乾番茜	olive oil
	dried parsley

Method
做法

1. 薯仔連皮洗淨，切件，浸於清水，去除過盛的澱粉黏液。
2. 用廚房紙印乾水分，放入碗內，加入鹽及橄欖油拌勻。
3. 放入焗爐以 180℃焗約 15 分鐘，翻轉再焗 5 分鐘或至金黃色，進食前灑上乾番茜更添風味。

1. Clean the skin of the potato. Cut into wedges. Soak in the water to remove the starch.
2. Wipe dry the potato with the kitchen paper. Put the potato in the bowl, then add the salt and olive oil and mix well.
3. Bake for 15 minutes at 180℃. Flip it over and bake for 5 minutes again until golden brown. Sprinkle with the dried parsley when serving.

Bento 2

Stir-fried Chicken Fillet with Carrot, Bell Pepper and Black Fungus

★

甘筍木耳
青椒炒雞柳

Ingredients
材料

急凍雞柳 3 條	3 pieces frozen chicken fillet
甘筍 1/3 碗	1/3 bowl carrot
乾木耳適量	dried black fungus
甜青椒 1/3 個	1/3 green bell pepper
洋葱 1/3 個	1/3 onion
蒜頭 2 瓣	2 cloves garlic
炒香白芝麻少許	stir-fried sesame

Marinade
醃料

生抽、薑汁、料理酒及麻油各少許　　light soy sauce
薑黃粉（可省略）　　　　　　　　　　ginger juice

cooking wine

sesame oil

turmeric powder (optional)

Seasonings
調味料

鹽 1/2 茶匙	1/2 tsp salt
生抽 1/2 茶匙	1/2 tsp light soy sauce
糖少許	sugar

Method
做法

1. 雞柳解凍，水加入鹽 2 湯匙，浸約 30 分鐘，略沖水及印乾，切條或切粒。
2. 雞肉加入醃料拌勻，醃約 30 分鐘。
3. 乾木耳浸發約 10 分鐘，切絲備用。
4. 青椒洗淨，去籽，切絲；甘筍去皮，切絲；洋葱切絲備用。
5. 油熱起鑊，放入雞肉炒香至兩面轉色，盛起。
6. 鑊內加入蒜片炒香，下洋葱絲拌炒，加入甘筍絲及木耳絲續炒，灑入調味料炒勻，再加入青椒絲拌勻，雞肉回鑊，熄火，灑上白芝麻。

1. Defrost the chicken fillet. Place in the water with 2 tbsp of salt and soak for 30 minutes. Rinse well and then cut into strips or dice, pat dry.
2. Mix the chicken fillet with the marinade and marinate for 30 minutes.
3. Soak the dried black fungus in water for 10 minutes until soft. Cut into shreds.
4. Clean the bell pepper. Remove the seeds and shred thinly. Peel the carrot and shred thinly. Cut the onion into shreds.
5. Stir-fry the chicken until done. Set aside.
6. Stir-fry the sliced garlic until fragrant. Add the onion and mix well. Put in carrot, black fungus and the seasonings. Add the bell pepper and stir well. Return the chicken lastly. Turn off the heat and sprinkle with the sesame.

午 餐 便 當 3 *一人分量

蛋白質：洋葱炒牛肉

蔬菜：西蘭花 / 椰菜花 / 福花、洋葱

自選常備菜：意式油漬彩椒（p.84）、和風淮山伴秋葵（p.89）

Skye's tips

> 紫色椰菜花可用白色椰菜花或西蘭花等
> 代替。

Bento 5

Stir-fried Beef with Onion

★

洋葱
炒牛肉

Ingredients
材料

牛肉 (分量為手掌般)
洋葱 1/4 個
蒜頭 2 瓣
紫色椰菜花 1/4 個

beef (the amount is same as the size of palm)
1/4 onion
2 cloves garlic
1/4 purple cauliflower

Marinade
醃料

生抽 2 茶匙　　2 tsp light soy sauce
糖少許　　　　sugar
料理酒適量　　cooking wine

Seasonings
調味料

鰹魚汁 3 湯匙 (按樽裝指示與適量水混和)
味醂少許

3 tbsp bonito sauce (follow the instruction and mix with enough water)
mirin

Method
做法

1. 牛肉用水沖洗，用廚房紙印乾水分，放入醃料拌勻。
2. 洋葱去皮、切絲；蒜頭拍扁，切粒備用。
3. 油熱起鑊，加入牛肉炒至 8 成熟，盛起。
4. 鑊內加入洋葱及蒜粒炒香，下鰹魚汁及味醂，最後下牛肉回鍋拌炒。
5. 椰菜花浸洗乾淨，加入橄欖油拌勻，以中小火隔水蒸 8-10 分鐘即成。

1. Rinse the beef and wipe dry with the kitchen paper. Mix with the marinade.
2. Remove the skin of the onion. Cut into shreds. Pat the garlic lightly and dice.
3. Stir-fry the beef until 80% done. Set aside.
4. Use the same wok. Stir-fry the onion and garlic until fragrant. Pour in the bonito sauce and mirin, mix well. Return the beef to the wok and stir well.
5. Cut the cauliflower into preferred size and soak in the water to remove the dirt. Drain well. Mix with olive oil and steam over medium-low heat for 8-10 minutes.

日常減醣便當

午餐便當 4 *一人分量

蛋白質：蒜蓉雞扒、玉子燒

蔬菜：秋葵、車厘茄

自選常備菜：醋漬紫洋蔥（p.82）

優質澱粉質：櫻花蝦紅米飯

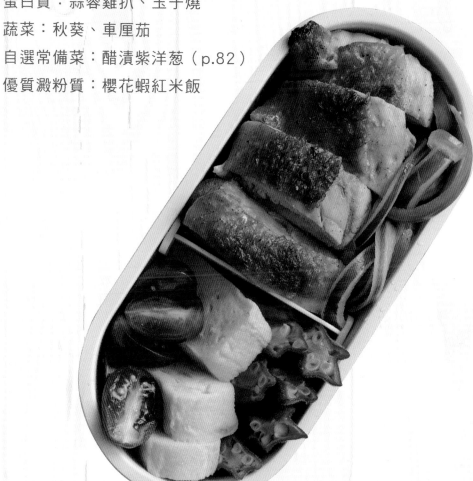

Skye's tips

如不喜歡秋葵的朋友，可嘗試配搭日式芥
辣及日式醬油伴秋葵，絕對有驚喜。

Bento 4

Garlic Chicken Fillet

★

蒜蓉雞扒

Method

做法

蒜蓉雞扒的醃法參考 p.40。雞扒放入已預熱焗爐，以 170-180℃焗 15 分鐘，翻轉再焗 6-8 分鐘或至熟（視乎雞扒大小及厚薄而定），取出。

Refer to p.40 for the marinating method. Put the chicken in the oven, bake for 15 minutes at 170-180°C. Flip it over and bake another 6-8 minutes or until done (depend on the size and thickness of the chicken). Set aside.

Bento 4

Dried Sakura Shrimps and Red Rice

★

櫻花蝦
紅米飯

Method

做法

做法參考 p.101，配雞扒做成便當，為優質澱粉質之選。

Refer to p.101 for the method. Brown rice is an ideal substitute for white rice and hence, is great for losing weight.

Bento 4

Tamagoyaki

★

玉子燒

Ingredients 材料

雞蛋 4 隻
水 40 毫升

4 eggs
40ml water

Seasonings 調味料

鰹魚汁 / 鰹魚粉適量
鹽及味醂各少許

bonito sauce or Japanese dashi

salt

mirin

Method 做法

1. 雞蛋拂勻，加入調味料及水拌勻。
2. 玉子燒鑊面掃上油，開小火，加入蛋漿均勻鋪滿表面，慢慢向同一方向捲起蛋皮，捲好後推至另一邊。
3. 於鑊面掃上油，準備做第二層，重複做至 5-6 層或至蛋漿用完，待涼後切件。

1. Whisk the eggs. Add the seasonings and water, mix well.
2. Spread a layer of oil in the tamagoyaki pan. Turn to low heat. Pour in the egg wash to spread over the pan. Roll the egg wash sheet slowly to the one side of the pan. Put the egg roll on the side.
3. Spread a layer of oil again. Repeat step 2 to make the second layer of the tamagoyaki. The tamagoyaki has 5 to 6 layers finally or until finish all the egg wash. Let it cool down and cut into pieces.

Bento 4

Okra and Cherry Tomatoes

★
秋葵、
車厘茄

Ingredients
材料

秋葵適量
車厘茄適量
okra
cherry tomatoes

Method
做法

1. 秋葵用鹽搓 30 秒，再用水沖淨，去掉蒂頭菱角邊緣表面。
2. 燒滾水加鹽，放入秋葵汆燙 1 分鐘，取出，立即放入冰水冰鎮以保持清爽口感。
3. 車厘茄洗淨，切半，放入飯盒。

1. Rub the okra with salt for 30 seconds. Rinse well and cut off the stem of the okra.
2. Bring the water to the boil. Put in the okra and boil for 1 minute. Take them out and soak in the ice water to keep the crunchy and fresh of the okra.
3. Rinse the cherry tomatoes and cut into halves. Put them in the lunch box.

健康有營的便當，
一家大小都適合！

BE
HEALTHY
eat
HEALTHY

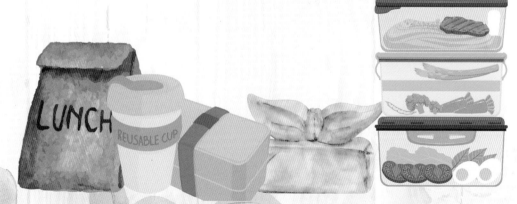

LUNCH

REUSABLE CUP

Chapter 5

Low-Carb Sets

日 常 減 醣 餐

Recipes

要做到輕鬆煮，不難！

記住以下的重點：

‧選優質蛋白質。

‧豐富多元蔬菜。

‧優質原形澱粉質。

只要選對食材，從中只作出微調，

減醣餐即變家常餐！

香茅豬扒併意式翠玉瓜麵

Lemongrass Pork Chops
Zoodles (Zucchini Noodles)

家常餐：可簡單配飯或各款麵食

★
香茅豬扒

Ingredients
材料

香茅豬扒（醃製方法參考 p.37）
車厘茄
自選沙律菜
煎蛋

lemongrass pork chops
(refer to p.37 for prepartion)
cherry tomatoes
salad greens of your choice
egg

Method
做法

1. 煮豬扒前，去掉豬扒表面過多的香茅醃料，以免表面過早煎成焦黑色。
2. 燒熱油鑊，放入豬扒以中大火煎封兩面，以免肉汁流失，加蓋，以小火續煎 10-12 分鐘，期間翻轉一次，上碟，配上沙律菜及煎蛋。
3. 或選用氣炸鍋，使豬扒上色更均勻。預熱 160℃氣炸 12-15 分鐘，期間翻轉，再調高溫度至 180℃氣炸 3-5 分鐘或至上色即可。

1. Before cooking the pork chop, it is recommended to remove excess lemongrass from the surface to prevent it from overcook.
2. If you prefer to pan-fry the pork chop, fry both sides of the pork chop over high heat then reduce to low heat and cover the pan. Continue cooking for 10-12 minutes, you can flip once or more until it is cooked. Serve with salad greens and fried egg.
3. You can choose to air-fry the pork chop. Preheat the air-fryer to 160℃, cook for 12-15 minutes, flip once. Then adjust the temperature to 180℃, air-fry for an additional 3-5 minutes or until golden brown. Adjust the cooking time according to the size and thickness of the pork chop.

意式翠玉瓜麵

Zoodles (Zucchini Noodles)

Ingredients
材料

意大利翠玉瓜或韓式翠玉瓜 2 條
蒜粒
車厘茄（可省略）
2 Italian zucchini or Korean zucchini
minced garlic
cherry tomatoes (optional)

Seasonings
調味料

鹽、黑椒碎及香草適量
salt, chopped black pepper and herbs

Method
做法

1. 翠玉瓜洗淨，用廚房紙抹乾，切去頭末兩段。
2. 用翠玉瓜麵專用刨器或刨刀，製成麵條狀備用。
3. 煎鍋內加入油及蒜粒炒香，加入翠玉瓜麵及車厘茄快炒 2 分鐘，以免炒煮時間太久致瓜麵出水。
4. 最後加入適量鹽、黑椒碎及自己喜好的香草調味，放於碟上享用。

1. Clean the zucchini and wipe dry with the kitchen paper. Cut off the head and the end of the zucchini.
2. Twist the zoodles with the hand-held spiralizer or the food cutter.
3. In the frying pan, add oil and minced garlic, stir-fry until fragrant. Add zoodles (zucchini noodles) and cherry tomatoes to the pan and stir-fry for 2 minutes .
4. Season with salt, black pepper and any preferred herbs.

Skye's tips

選用意大利翠玉瓜、韓式翠玉瓜及中式翠
玉瓜都能成功製成 Zoodles；但由於青瓜
水分較多，不適合扭出麵條狀。

山楂咕嚕雞肉丸子、蝦米炒椰菜

Sweet and Sour Chicken Meatballs
Stir-fried Cabbage with Dried Shrimps

減醣餐：糙米藜麥小米飯

家常餐：白飯（建議平常可加小米、燕麥等與白米同煮）

山楂咕嚕
雞肉丸子

Sweet and Sour Chicken Meatballs

Ingredients

材料

丸子材料

雞胸或雞柳約 600 克

甘筍 1/3 條

粟米 1/2 條（切粒）

洋葱 1/4 個

Ingredients for chicken meatballs

600 g chicken breast or fillet

1/3 carrot

1/2 cob of corn (diced)

1/4 onion

配菜材料

洋葱 1/2 個

三色椒各 1/2 個（切件）

菠蘿適量（切件）

蘋果 1/2 個（切件）

薑片及蒜粒各適量

Ingredients for side veggies

1/2 onion

each of 1/2 bell peppers (red, green and yellow; cut into pieces)

pineapple (cut into pieces)

1/2 apple (cut into pieces)

sliced ginger

diced garlic

Seasonings
調味料

丸子調味料

鹽 2 茶匙

糖 2 茶匙

生抽 2 湯匙

麻油 1 湯匙

料理酒 2 茶匙

油及粟粉各適量（後下）

Meatball seasonings

2 tsp salt

2 tsp sugar

2 tbsp light soy sauce

1 tbsp sesame oil

2 tsp cooking wine

oil (add at last)

corn starch (add at last)

自製咕嚕汁

山楂 1 碗

冰糖 1/2 碗

（＊可另加適量茄汁、白醋調味）

Homemade sweet and sour sauce

1 bowl dried hawthorn

1/2 bowl rock sugar

(you can add tomato sauce and white vinegar for seasoning to taste, optional)

120

Method
做法

1. 山楂洗淨，放於滾水煲 20 分鐘，加入冰糖煮溶，過篩，用大湯匙壓出山楂汁，盛起備用。

2. 甘筍、粟米及洋葱切好放入攪拌機；雞柳解凍，用水沖洗，以廚房紙吸乾水分，切件，同放入攪拌機，加入調味料攪拌成膠狀，最後加入生粉及油輕拌。

3. 預熱小丸子烤盤，掃油後逐一加入雞肉，慢慢推成丸子狀。如沒有丸子烤盤，可用手搓成肉丸或漢堡扒，以平底鑊煮熟即可。

4. 燒熱平底鑊，放入薑片、蒜粒及洋葱炒香，加入三色椒略炒。

5. 山楂汁可隨喜好加入茄汁及白醋調味，放入煮至收汁，下丸子拌至掛汁，最後下蘋果及菠蘿炒勻，上碟。

1. Wash the dried hawthorn and boil them in water for about 20 minutes. Add rock sugar and boil until the sugar is dissolved. Strain and press the dried hawthorn to extract the juice. Transfer the juice to a large bowl and set aside.

2. Put the carrot, corn and onion in the blender. Rinse the chicken and wipe dry with kitchen paper. Cut the chicken into pieces, put it in a blender. Add the seasonings and blend together. Add corn starch and oil at last, gently mix until the mixture becomes sticky.

3. Preheat a meatball cooking pan and brush it with oil. Place the chicken mixture into the pan and shape it into meatballs with a spoon. If you do not have a meatball baking pan, you can shape the mixture into meatballs by hand. Alternatively, you can shape them into burger patties or other desired forms and cook them in a frying pan.

4. Stir-fry ginger, garlic and onion until fragrant. Stir-fry the bell peppers briefly.

5. Add the sweet and sour sauce to the frying pan, you can add tomato sauce and vinegar according to your preference. Cook until it reaches the desired consistency. Add the apple and pineapple and stir-fry. Put the meatballs in the pan and stir-fry until they are coated with the sauce. Serve.

Stir-fried Cabbage with dried shrimps

★

蝦米炒椰菜

Ingredients
材料

椰菜 1/2 個
蝦米 1/2 碗
乾葱 2 粒
蒜頭 3 瓣

1/2 cabbage
1/2 bowl dried shrimps
2 shallots
3 cloves garlic

Seasonings
調味料

鹽約 1 1/2 茶匙
糖少許
麻油 1 湯匙

1 1/2 tsp salt
sugar
1 tbsp sesame oil

Method
做法

1. 椰菜洗淨，切塊，瀝乾水分備用。
2. 蒜頭去皮、拍扁，切片；乾葱去衣，切粒。
3. 燒熱油，下蝦米、蒜片及乾葱爆香，加入椰菜炒勻，灑入鹽、少許糖及麻油，上碟。

1 Rinse the cabbage and cut into pieces. Drain well.
2 Remove the skin of garlic. Pat lightly and slice. Peel the garlic and dice.
3 Stir-fry the dried shrimps, garlic and shallot until fragrant. Add the cabbage and stir well. Sprinkle with salt, a pinch of sugar and sesame oil. Stir well and serve.

Quinoa, Millet and Red Brown Rice

★ 紅糙米藜麥小米飯

Ingredients
材料

藜麥（可選三色藜麥）
小米
糙米
紅糙米
白米（可省略）
（＊可按人數調整各款米的比例，可隨意混合）

quinoa
millet
brown rice
red brown rice
white rice (optional)
(*adjust the ratio of the rice according to the number of people or your preference)

Method
做法

1. 糙米及紅糙米用水浸泡約 3 小時，瀝乾水分。
2. 小米、藜麥及白米洗淨，與已浸泡糙米及紅糙米同放入電飯煲，以一比一的水量（與白飯水量一樣），按煮飯模式即可。

1. Soak the brown rice and red rice in the water for 3 hours. Drain well.
2. Rinse the millet, quinoa and white rice. Cook them with brown rice and red brown rice using the regular rice cooking mode. The ratio of rice and water is 1:1.

Skye's tips

印度米（Basmati rice）升糖值比白米低，是關注血糖水平人士食用白米的替代選擇，同時保留吃米飯的咀嚼感，如購買到 Brown basmati rice 印度糙米更理想。煮印度米需要比白米更多的水，大約 1 份米：1 3/4 份水，可加橄欖油、香草或薑黃粉等同煮，更添風味。

鹽酥雞、魷魚乾蝦米炒雜菜

Crispy Popcorn Chicken
Stir-fried Vegetables with Dried Squid and Dried Shrimps

家常餐：建議雜菜拌入米粉同炒

鹽酥雞

Crispy Popcorn Chicken

★

Ingredients
材料

雞扒 / 雞腿 / 雞胸 2 塊	2 pieces chicken fillets/chicken breast
九層塔 1 棵	1 sprig basil
青檸 1 個	1 lime
雞蛋 1 隻	1 egg

Marinade
醃料

蒜蓉 1 茶匙	1 tsp minced garlic
五香粉 1/4 茶匙	1/4 tsp five-spices powder
薑黃粉 1/4 茶匙	1/4 tsp turmeric powder
糖 1 茶匙	1 tsp sugar
生抽 2 湯匙	2 tbsp light soy sauce
胡椒粉適量	ground white pepper
料理酒 1 湯匙	1 tbsp cooking wine
麻油 1 茶匙	1 tsp sesame oil
檸檬汁 2 茶匙	2 tsp lemon juice
鹽 1 茶匙	1 tsp salt

Deep-frying batter
炸粉料

燕麥片 1 碗	1 bowl rolled oats
白芝麻 1/2 碗	1/2 bowl white sesames

Method
做法

1. 雞件略沖，瀝乾水分，切件，與醃料拌勻後，加入雞蛋拌勻。
2. 白芝麻及燕麥打成粉，備用。
3. 雞件均勻地沾滿芝麻燕麥粉，醃一會。
4. 雞件放入熱油炸至金黃色，約 5 分鐘（視乎雞件大小）。
5. 九層塔可放入熱油炸 30-60 秒至香脆，伴於鹽酥雞，進食時灑上調味料及青檸汁。

1. Rinse and wipe dry the chicken, then cut the chicken into bite-sized pieces. In a large bowl, add the marinade and chicken, mix well then add an egg, continue to mix.

2. Put the rolled oats and white sesames in the blender and blend together.

3. Once the chicken has marinated, add the blended sesame and oats powder directly to the chicken. Toss the chicken in the powder until evenly coated.

4. Deep-fry the chicken in small batches until golden brown (about 5 minutes depending on the size of the chicken).

5. You can deep fry the basil as well, about 30-60 seconds. Remove the basil and add to the fried chicken. Sprinkle with the seasoning and lime juice to taste.

★

魷魚乾蝦米炒雜菜

Stir-fried Vegetables with Dried Squid and Dried Shrimps

Ingredients
材料

腩頭瘦肉 1 碗	1 bowl of port shoulder butt
椰菜 1/3 個	1/3 cabbage
洋葱 1/2 個	1/2 onion
甘筍 1/2 個	1/2 carrot
銀芽 1 斤（600 克）	600 g mungbean sprouts
蝦米 1/3 碗	1/3 bowl dried shrimps
魷魚乾 1 塊	1 piece dried squid
薑 3-4 片	3-4 slices ginger
蒜頭 4 瓣	4 cloves garlic
葱 2 條	2 sprigs spring onion
炒香白芝麻適量	stir-fried sesames

Seasonings
調味料

鹽麴 2 茶匙（做法參考 p.74）
糖 1 茶匙
麻油 1 茶匙
料理酒 1 茶匙
蠔油 1 茶匙（可省略）
蒜粉 1 茶匙（可省略）

2 tsp shiokouji
(refer to p.74 for the method)
1 tsp sugar
1 tsp sesame oil
1 tsp cooking wine
1 tsp oyster sauce (optional)
1 tsp garlic powder (optional)

Marinade
腩頭肉醃料

鹽 2 茶匙	2 tsp salt
生抽 1 茶匙	1 tsp light soy sauce
糖 1/2 茶匙	1/2 tsp sugar
麻油 1 茶匙	1 tsp sesame oil

Method
做法

1. 胸頭瘦肉切條，與醃料拌勻，備用。
2. 椰菜洗淨，瀝乾水分，切絲備用。
3. 洋葱去皮，切幼條；甘筍去皮，切絲；銀芽洗淨，瀝乾備用。
4. 蝦米洗淨，用水浸約 20 分鐘，備用。
5. 魷魚乾沖淨，用水浸泡後，去掉透明骨，切幼條備用。
6. 薑洗淨，連皮切絲；蒜頭去皮、拍扁，切粒；葱白切絲，青葱切粒。
7. 油熱起鑊，下薑及蒜粒炒香，加入蝦米、肉絲及魷魚乾拌炒，盛起備用。
8. 原鑊加入椰菜、甘筍絲、銀芽及調味料炒勻，將蝦米肉絲回鑊拌炒，進食時灑上白芝麻即成。

1 Cut the pork into shreds. Mix with the marinade and set aside.
2 Rinse the cabbage and drain well. Finely shred.
3 Remove the skin of onion and cut into thin strips. Peel the carrot and shred. Rinse the mungbean sprouts and drain.
4 Rinse the dried shrimps. Soak them in the water for 20 minutes.
5 Clean the dried squids and soak in the water. Remove the inner bone and cut into thin strips.
6 Clean the ginger and cut into pieces with the skin on. Remove the skin of garlic, pat lightly and dice. Shred the white part of spring onion and dice the green part.
7 Stir-fry the ginger and garlic until fragrant. Add the dried shrimps, pork shreds and dried squid and stir-fry. Set aside.
8 In the same wok, put in the cabbage, carrot, mungbean sprouts and seasonings, mix well. Return the dried shrimps and pork to the wok. Sprinkle with the sesames when serving.

Skye's tips

如時間許可，建議將雞蛋煎成蛋皮後切絲，配搭其他食材更顯風味。

番茄蝦湯、田園溫野菜拼盤

Tomato Shrimp Soup
Warm Vegetable Platter

家常餐：建議用番茄蝦湯做成湯意粉或湯通粉

Tomato Shrimp Soup

★

番茄蝦湯

Ingredients
材料

新鮮蝦 1 斤（600 克）
番茄 5 個
洋葱 1 個

600 g fresh shrimps

5 tomatoes

1 onion

Seasonings
調味料

味醂適量
香料適量

mirin

herbs

Seasonings for shrimps broth
蝦湯調味料

鹽 2 茶匙	2 tsp salt
蒜粉 1 茶匙	1 tsp garlic powder
胡椒粉適量	ground white pepper
料理酒 1 茶匙	1 tsp cooking wine
紅椒粉 2 茶匙	2 tsp paprika

Method
做法

蝦湯做法

1. 新鮮蝦洗淨，用牙籤挑走蝦腸，去掉蝦頭及殼備用，蝦肉放於另一碗。
2. 油熱起鑊，蝦頭及蝦殼略印乾水分，下鑊炒香，期間可按個人口味加入鹽、蒜粉、胡椒粉及紅椒粉，於鑊邊灒少量料理酒。
3. 加入已煮沸滾水，加蓋煮 10 分鐘，熄火，加蓋繼續待 20 分鐘，用隔篩濾出蝦湯，蝦殼保留（用於自製蝦粉，見 p.134）。

Method for the shrimps broth

1 Rinse the fresh shrimps. Take out the veins with the toothpick. Remove the heads and the shells of the shrimp, keep them for later use. Put the shrimp meat in the bowl.

2 Wipe dry the shrimp head and shells. Stir-fry until fragrant. Season with the salt, garlic powder, pepper and paprika to taste. Add the cooking wine to taste.

3 Pour the boiling water and cover the lid to cook for 10 minutes. Turn off the heat and keep it for 20 minutes. Strain the shrimp soup and reserve the shell for making the shrimp powder (refer to p.134).

番茄湯做法

1. 番茄洗淨，去皮、切件；洋蔥去皮，切幼條。
2. 油熱起鑊，加入洋蔥略爆炒，待釋出洋蔥香，加味醂調味。
3. 加入番茄炒勻（先不用加水），待番茄釋出水分，加蓋煎煮約 10 分鐘，加入蝦湯，試味，加入其他喜好的香料或調味料。
4. 如用攪拌棒打成濃湯，盛起部分洋蔥及番茄蓉，打好濃湯後放回鍋內。
5. 蝦肉煎香、蒸或灼，最後放入蝦湯同煮 5-8 分鐘，享用。

Method for the tomato soup

1 Wash and peel the tomatoes, cut into pieces. Peel the onion and cut into thin strips.

2 Stir-fry the onion until fragrant. Add mirin or additional seasonings according to taste.

3 Add the tomatoes and stir-fry till the tomato juice come out. It is without adding water. Cover and simmer for about 10 minutes. Add the shrimp broth and adjust the flavour by adding preferred herbs or seasonings.

4 If you prefer a creamy soup, you can use a blender to puree a portion of the onion and tomato mixture before returning it to the pan.

5 Shrimp can be sautéed, steamed or blanched separately. The most convenient method is to add them directly to the soup and cook for 5-8 minutes.

Skye's tips

每次購買新鮮蝦，除了白灼、蒸或煎蝦碌外，平日在家煮家常餐，還可以利用蝦殼的鮮味熬出濃郁香味的蝦湯或蝦汁伴意粉。

Warm Vegetable Platter

田園溫野菜拼盤

Ingredients 材料

Seasonings 調味料

＊ 隨個人喜好任選 3-5 款，分量隨意

翠玉瓜 / 意式青瓜
茄子
三色甜椒
西蘭花
甘筍
薯仔
洋葱
白蘑菇
蒜頭

＊ Choose 3-5 types according to personal preference, the amount depend on your need

zucchini
eggplant
bell peppers
broccoli
carrot
potato
onion
white button mushrooms
garlic

＊ 按個人口味調節，毋須固定分量

牛至
百里香
罌粟籽或大麻籽（可省略）
橄欖油
生抽
蜜糖
檸檬汁

＊ The amount depends on your preference

oregano
thyme
poppy seeds or hemp seeds（optional）
olive oil
light soy sauce
honey
lemon juice

Method
做法

1. 翠玉瓜／意式青瓜及茄子洗淨，抹乾水分，切細件，放於大碗，灑入鹽拌勻待 10 分鐘，瀝乾水分備用。
2. 三色椒洗淨，去籽，切細件；紅蘿蔔去皮，切細件。
3. 薯仔去皮，切件，用水浸 10 分鐘，濾走水分備用。
4. 洋葱去皮，切件；蘑菇切件備用。
5. 西蘭花洗淨，切件，灼熟備用。
6. 調味料放入大碗調勻，放入所有蔬菜拌勻後（西蘭花除外），排於已塗油的焗盤，預熱焗爐 180℃。
7. 蒜頭頂部整個切去，放入焗盤，蒜頭面灑上適量橄欖油、鹽及黑椒碎，焗約 40 分鐘；出爐後加入西蘭花拌勻即成，伴番茄蝦湯享用。

1. Clean the zucchini and eggplant. Wipe dry and cut them into thin pieces. Place them in a large bowl. Add salt and mix well, let them sit for 10 minutes. Drain the excess water and set aside.
2. Clean the bell peppers, remove the seeds and cut them into pieces. Peel the carrot and cut into thin pieces.
3. Peel the potato and cut into pieces. Place them in a bowl, soak in water for 10 minutes, and then drain.
4. Peel the onion and cut into pieces. Cut the mushrooms into pieces and set aside.
5. Rinse the broccoli and cut it into small florets, put in the boiling water and bring to boil , take out and set aside.
6. Prepare another large bowl for mixing the sauce. Add the seasonings and mix well. Add all the vegetables (except the broccoli) and mix well. Brush an ovenproof dish with oil and place in the vegetables. Preheat the oven to 180℃ .
7. Cut the top of the garlic cloves and place in the baking dish, drizzle with olive oil, salt, and black pepper. Bake the vegetables in the oven for about 40 minutes until golden and completely soft. After removing from the oven, add the blanched broccoli and mix well.

★
自製蝦粉

如家裏有較強力的攪拌機,非常建議大家試試自家製蝦粉,平日可用於炒菜、餃子餡調味及加添湯內帶出海鮮味,用法與大地魚粉及香菇粉相似。

Ingredients
材料

新鮮蝦頭及蝦殼
fresh shrimp heads and shells

Method
做法

1. 蝦頭及蝦殼鋪平焗盤,預熱焗爐 160℃ 焗 40 分鐘。
2. 蝦殼焗至乾身後,放室內待涼,放入食物處理器(Food Processor)或攪拌機打成粉,期間可加鹽或蒜鹽製成蒜味蝦粉等不同口味。
3. 蝦粉打好後,放入已清潔消毒及抹乾的玻璃瓶儲存。

1. Lay the shrimp shells flat on a baking tray and preheat the oven to 160℃ for 40 minutes.
2. Let them cool at the room temperature. Place them in a food processor or blender and process into powder. You can add salt or garlic salt to create different flavours of shrimp powder.
3. Store the shrimp powder in a clean and sterilized glass container or jar.

Skye's tips

蝦頭含不少蝦膏及蝦肉,平常我會先剪碎再炒,讓蝦湯變得更濃郁、更惹味。

滷肉滷蛋伴綠豆粉絲

Braised Pork and Braised Eggs with Mungbean Vermicelli

家常餐：配上白飯，即成最受小朋友歡迎的碟頭飯

Ingredients
材料

五花腩 1/2 斤（300 克，與胛頭肉的
比例是 1：3）
胛頭瘦肉 1 1/2 斤（900 克）
綠豆粉絲 2 束
花椒 3 湯匙
八角約 10 粒
香葉 4 塊
洋蔥 1 個
薑 5 片
蒜 1/2 個
乾蔥 2 粒
櫻花蝦、乾秀珍菇（可隨意加添）
300 g pork belly (the ratio with pork
shoulder butt is 1:3)
900 g pork shoulder butt (lean)
2 bundles mungbean vermicelli
3 tbsp Sichuan peppercorns
10 pieces star anise
4 bay leaves
1 onion
5 slices ginger
1/2 garlic
2 shallots
dried sakura shrimp (optional)
dried oyster mushrooms (optional)

Seasonings
調味料

＊按豬肉分量調整調味
冰糖約 50 克
生抽 120 毫升
老抽 4 湯匙
料理酒 100 毫升
蠔油 2 湯匙
麻油 1 湯匙
**＊Adjust the seasonings
according to the amount of the meat**
50 g rock sugar
120ml light soy sauce
4 tbsp dark soy sauce
100ml cooking wine
2 tbsp oyster sauce
1 tbsp sesame oil

Method

做法

1. 五花腩及胸頭瘦肉洗淨，切成幼條狀。
2. 薑洗淨，切片（可去皮，或洗淨薑皮保留）。
3. 洋葱及乾葱切粒；蒜頭去皮，拍扁備用。
4. 花椒及八角用茶袋盛好，備用。
5. 油熱起鑊，加入五花肉條及胸頭肉條，灑入少許鹽炒香，待表面大致煮熟，盛起。
6. 原鑊保留，放入薑片、乾葱、洋葱及蒜頭炒香（此時可加入櫻花蝦及浸軟秀珍菇拌炒，加添香味，提升層次）。
7. 肉條回鑊，鑊中間留位，加入生抽、老抽及糖拌至糖溶化，炒勻，於鑊邊灒入料理酒炒香。
8. 加水蓋面，放入花椒八角茶袋及香葉，加蓋，以中大火煮 20 分鐘，轉小火，試味後隨個人喜好加入蠔油、麻油或生抽。
9. 雞蛋烚熟，去殼，放進滷肉鍋同浸，加蓋，以慢火燜煮 1 小時，熄火（不開蓋）。
10. 進食前以大火煮至收汁，最後伴已灼熟的綠豆粉絲，上碟。

1. Clean and dry the meat. Cut the pork into thin strips.
2. Clean the ginger, cut into pieces.
3. Finely chop the onion and shallots. Remove the skin of the garlic and pat lightly.
4. Place Sichuan peppercorns and star anise in a tea bag.
5. Stir-fry the meat strips and sprinkle with a pinch of salt until lightly browned and transparent. Set aside.
6. Add ginger slices, shallots, chopped onion and garlic in the same wok. Stir-fry until fragrant. For an enhanced flavour, you can add sakura shrimps and soaked dried oyster mushrooms and stir-fry together.
7. Return the meat to the wok, leaving a hole in the center. Add light soy sauce, dark soy sauce and sugar until the sugar is dissolved. Stir-fry until the sauce coats the meat evenly. Pour cooking wine along the edge of the wok and stir-fry to release the aroma.

8. Add water to cover the ingredients, then add the tea bag containing Sichuan peppercorns and star anise, as well as the bay leaves. Cover and simmer over medium-high heat for 20 minutes. Reduce to the low heat. Taste the sauce and adjust the flavour by adding oyster sauce, sesame oil, or light soy sauce according to personal preference.

9. Add the boiled eggs to the pot, allowing them to soak in the sauce. Cover the pot and simmer over low heat for 1 hour. Keep the lid on after turning off the heat.

10. Before serving, cook over high heat to reduce the sauce until it thickens to your desired consistency. Serve with the cooked mungbean vermicelli.

Skye's tips

· 飽住瘦食譜建議：五花肉與腸頭肉比例 1:5；一般食譜建議：五花肉與腸頭肉比例 1:3；油脂豐富比例 1:1。
· 天冷時，滷肉放室溫數小時後，表面會呈一層油脂，開火翻熱時可先掏起白色油脂層，免卻進食過多脂肪。
· 減醣期間建議進食適當天然油脂，只要聰明地選擇，可以有效享用美食時不吸收多餘脂肪。

粟米蘑菇湯、燒排骨伴椰菜花

Corn and Mushroom Soup
Grilled Spareribs and Cauliflower

家常餐：配搭飯、意粉同樣方便美味

★
粟米
蘑菇湯

Ingredients
材料

Seasonings
調味料

粟米 2 條
洋葱 1/2 個
白蘑菇 1 碗
2 cobs of corn
1/2 onion
1 bowl white button mushrooms

鹽 1/2 茶匙
糖 1/2 茶匙
味醂適量（可省略）
胡椒粉或黑胡椒粉適量
1/2 tsp salt
1/2 tsp sugar
mirin (optional)
ground white pepper /ground black pepper

Skye's tips

· 這是非常簡易的家常煮法，有時間或想多點西餐原味
  的話，可用忌廉及麵粉的白汁製作方法。
· 這款仿西式的家常粟米湯，除可用作西餐湯，可變身
  粟米肉粒飯的汁（需要煮得濃稠一點），也可做成小
  朋友喜愛的粟米湯通粉。如家裏有矽膠冰模具，分量
  可煮多些，倒入模具冷凍，煮食時直接加入粟米冰粒
  加熱就可以了。

Method
做法

1. 粟米洗淨，切出粟米粒；洋葱去衣，切粒；白蘑菇抹淨，切粒。
2. 油熱起鑊，加入白蘑菇炒香，盛起備用（如想增添風味，可偶爾用牛油炒蘑菇）。
3. 鍋內燒熱油，加入洋葱炒香，加少許鹽、糖、味醂及胡椒調味，加入粟米粒拌勻，注入沸水剛蓋過食材面，加蓋，以中火煮 15 分鐘（可加粟米芯同煮，隨後取出）。
4. 可準備手提攪拌棒或攪拌機攪打粟米湯，先盛起適量粟米粒及洋葱，不要全部打爛，可有粟米粒及粟米蓉不同的口感。
5. 打好的粟米湯，可隨意加入牛奶或植物奶享用。

1. Rinse the corn and break the corn kernels from the cob. Remove the skin of onion and dice. Clean the button mushroom with the kitchen paper and dice.
2. Stir-fry the button mushrooms until fragrant. You can occasionally use butter to enhance the flavour.
3. Add the onion to the pan and stir-fry until fragrant. Season with a little of salt, sugar, mirin and pepper. Add the corn kernels and mix well. Add enough boiling water just to cover the ingredients. Cover the pot and simmer over medium heat for 15 minutes (can add the corn cob for boiling, then take out).
4. Prepare a handheld blender or a regular blender to puree the corn soup. Scoop out a suitable amount of corn and onion before blending. Do not blend everything at once to retain a mix of corn kernels and corn puree for different textures.
5. After blending the corn soup, you can add milk or plant-based milk according to your preference.

Grilled Spareribs and Cauliflower

★ 燒排骨伴椰菜花

Method
做法

1. 燒排骨醃製法參考 p.43。
2. 預熱焗爐 220℃，放入排骨烤焗 25 分鐘，翻轉再焗 5-8 分鐘，可塗抹蜜糖再焗 2 分鐘，配上已灼各式椰菜花，即可享用。

1. Refer to p.43 for the marinating method.
2. Preheat the oven at 220℃. Put the spareribs in the oven and grill for 25 minutes. Flip once and grill 5-8 minutes again. You can apply honey on top, bake for 2 minutes. Serve with the cooked cauliflower.

Skye's tips

- 基本上，這道燒排骨可配上任何蔬菜，適合自家西餐，更可於雪櫃取出常備菜當配菜。
- 豬肋骨可整排放入焗爐，亦可預先切成一條條烤焗，更惹味。

和風三文魚親子丼、大葱炒雜菇

Japanese Salmon and Salmon Roe Don
Stir-fried Assorted Mushrooms with Peking Scallions

減醣餐： 椰菜花飯

家常餐： 配搭白飯或紫菜飯糰

Ingredients
材料

＊一碗分量

三文魚柳 1 條
紫洋葱 1/4 個
雞蛋 2 隻
椰菜花 1/3 個
三文魚子 / 飛魚子適量
葱適量

＊1 bowl for serving

1 piece salmon fillet
1/4 purple onion
2 eggs
1/3 head of cauliflower
salmon roe or flying roe
spring onion

Sauce
湯汁調味料

＊拌勻

口式高湯 / 昆布湯 / 鰹魚汁約 80-100 毫升（濃縮版請按包裝指示用水以適當比例調開）
日式醬油 / 生抽 1 1/2 湯匙
味醂 1 1/2 湯匙

＊ mixed well

80-100ml Japanese stock/kelp stock/bonito sauce (follow the instruction to dilute the sauce with enough water)
1 1/2 tbsp Japanese soy sauce/light soy sauce
1 1/2 tbsp mirin

Method
做法

1. 燒熱油，三文魚柳放入平底鑊煎香兩面，加蓋，以小火焗煮 6-8 分鐘，或放入氣炸鍋以 150℃炸 8 分鐘。
2. 葱切成葱白及青葱兩部分，洗淨備用。
3. 在小型圓形平底鑊加油，油熱起鑊下洋葱炒香，加入湯汁調味料煮至透明狀，放入三文魚扒於中間，加入味醂。
4. 雞蛋輕輕拂勻（看見蛋白及蛋黃）。蛋液分兩次慢慢倒入，灑上葱白，以小火煮約 2 分鐘或至雞蛋凝固，熄火，加蓋焗 2 分鐘。
5. 椰菜花浸洗乾淨，瀝乾水分，以刨刀刨絲或用刀切碎。
6. 燒熱油，加入椰菜花碎、鹽少許及胡椒粉拌炒，期間可逐少加入水，熟透後放於丼飯碗內。
7. 雞蛋及三文魚輕輕地放於椰菜花飯面，加三文魚子／飛魚子，灑上青葱即成。

1. Fry the salmon fillet until both sides are golden. Cover the lid and cook over low heat for 6-8 minutes. Or you can air-fry the salmon fillet at 150℃ for 8 minutes.
2. Clean and cut the spring onion into white part and green part, set aside.
3. Stir-fry the onion until fragrant. Pour the sauce and cook the onion to transparent. Place the salmon fillet in the center of the sauce. Add the mirin.
4. Whisk the egg lightly (egg yolk and egg white separately). Pour the egg wash in the small pan at two times. Sprinkle with the white part of the spring onion, cook over low heat for 2 minutes or until the egg set. Turn off the heat. Cover the lid and let it sit for 2 minutes.
5. Soak the cauliflower in the water, drain well. Grate into shreds or finely chop the cauliflower.
6. Stir-fry the chopped cauliflower, add a pinch of salt and ground white pepper. Stir-fry for a while and pour a little of water each time. Put the cauliflower rice in the Don bowl when it is done.
7. Put the salmon and egg on the top of cauliflower. Decorate with the salmon roe or flying roe. Sprinkle with the green part of spring onion to decorate.

大葱炒雜菇

Stir-Fried Assorted Mushrooms with Peking Scallion

Ingredients
材料

大葱 1 條
鴻禧菇、秀珍菇、杏鮑菇（自行挑選）
薑 2 片
蒜頭 2 瓣
1 sprig Peking scallion
Shiro-shimeji mushrooms, oyster mushrooms,
king oyster mushrooms (choose as you like.)
2 slices ginger
2 cloves garlic

Seasonings
調味料

鹽或鹽麴少許
糖少許
胡椒粉或黑胡椒粉
salt or shiokouji
sugar
ground white pepper or
black pepper

Method
做法

1. 大葱洗淨，斜切成段。
2. 隨心選擇自己喜好的菇菌，切條、切段或撕成塊狀，備用。
3. 油熱起鑊，加入薑及蒜肉炒香，加入菇菌，略加調味料拌炒均勻即成。

1. Rinse the scallion well and cut into pieces diagonally.
2. Choose the mushrooms you like. Cut into strips or sections or tear into pieces.
3. Stir-fry ginger and garlic until fragrant. Put in the mushrooms and mix well. Sprinkle the seasonings and stir-fry. Serve.

蔬果汁牛仔骨、黑豆牛油果藜麥沙律

Baked Beef Short Ribs with Fruit Juice Avocado, Black Beans and Quinoa Salad

家常餐：可配搭拉麵享用

Avocado, Black Beans and Quinoa Salad

★ 黑豆牛油果藜麥沙律

 Ingredients
材料

沙律三色椒
番茄
罐裝黑豆
罐裝紅腰豆
牛油果
藜麥

bell pepper (red, green and yellow, for salad)
tomatoes
canned black beans
canned kidney red beans
avocado
quinoa

 Salad dressing
沙律汁

青檸汁
鮮榨橙汁（連橙肉）
黃芥末醬
橄欖油
鹽
黑椒碎
香草

lime juice
fresh orange juice (with flesh)
mustard
olive oil
salt
chopped black pepper
herbs

Skye's tips

沙律汁除用青檸外，可加入百香果或其他橘子類水果等，同樣清香開胃。

Method
做法

1. 三色椒洗淨，去籽，切小塊；番茄洗淨，切小塊。
2. 罐裝黑豆及紅腰豆瀝乾水分。
3. 藜麥用水浸軟，煮熟。
4. 將全部材料置於碟上，淋上拌勻的沙律汁享用。

1. Rinse the bell peppers. Remove the seeds and cut into small pieces. Rinse the tomatoes and cut into small pieces.
2. Drain the canned black beans and kidney red beans.
3. Soak the quinoa in the water and cook until done.
4. Place all the ingredients in the plate and pour the salad dressing on the top when serving.

★
蔬果汁
牛仔骨

Method
做法

蔬果汁牛仔骨醃製法參考 p.44，放入平底鑊煎熟，或以 180℃氣炸 5 分鐘，翻轉另一面再炸 5 分鐘，伴藜麥沙律享用。

Refer to p.44 for the marinating method. Put the beef short ribs in the pan and fry until done. Or you can air-fry the ribs at 180℃ for 5 minutes. Flip once and fry 5 more minutes. Serve with quinoa salad.

銀鱈魚西京燒、欖油炒蘆筍及金菇、熟番薯、意式油漬彩椒

Grilled Saikyo Miso Black Cod
Stir-fried Asparagus and Enoki Mushroom with Olive Oil
Cooked Sweet Potato
Italian-style Oil-pickled Bell Peppers

家常餐：建議配蕎麥麵

Grilled Saikyo Miso Black Cod

銀鱈魚
西京燒

Ingredients
材料

銀鱈魚 4 塊	4 pieces black cod
味噌 2-3 湯匙	2-3 tbsp miso
清酒 / 米酒 2 湯匙	2 tbsp sake/rice wine
味醂 3 湯匙	3 tbsp mirin
薑 4 片	4 slices ginger

Method
做法

1. 銀鱈魚解凍，灑入鹽待 20 分鐘，略沖水後，用廚房紙印乾水分。
2. 碗內加入味噌、味醂及清酒拌勻，試味。
3. 銀鱈魚及調味料放入儲存盒，均勻地塗抹兩面，加蓋冷藏，翌日使用。
4. 煎煮前用廚房紙略印魚身表面，以防容易燒燶。銀鱈魚放入平底鑊煎熟，或用氣炸鍋 160℃烤 10 分鐘，期間可翻面一次。

1. Defrost the cod. Sprinkle with salt and let it sit for 20 minutes. Rinse then wipe dry with the kitchen paper.
2. Mix the miso, mirin and sake in the bowl. Try and adjust the taste.
3. Put the cod and the seasonings in the food container. Rub the seasonings on the both sides of the cod, cover the lid and regfrigerate.
4. Wipe dry the cod surface before cooking. Fry in the pan or air-fry at 160℃ for 10 minutes. Flip once during the cooking.

日常減醣餐

151

★

欖油炒蘆筍及金菇

Stir-fried Asparagus and Enoki Mushroom with Olive Oil

Method
做法

1. 蘆筍沖淨，瀝乾水分。
2. 金菇切掉末端，與蘆筍同放入鑊內，用橄欖油及鹽略炒，盛起金菇，加入水 2 湯匙，加蓋，煮蘆筍數分鐘，以意式油漬彩椒伴碟。

1. Clean the asparagus and Enoki mushrooms.
2. Stir-fry with olive oil and salt. Put the Enoki mushrooms on the plate. Add 2 tbsp of water, cover the lid to cook the asparagus for few more minutes. Serve with the pickled bell peppers or tsukemono of your choice.

★

熟番薯

Cooked Sweet Potato

Method
做法

番薯洗淨，放入熱水內焗熟或蒸熟，上碟，成為優質澱粉質之選。

Clean the sweet potatoes. Boil in the water or steam until done. Or pair with any less processed carbs.

Skye's tips

一次可多醃幾片魚作為 meal prep 其中一款蛋白質，可試換成三文魚或鯖魚。

椰菜肉片鍋、木魚碎冷豆腐

Sliced Pork and Cabbage Hot Pot
Cold Tofu with Bonito Flakes

家常餐：建議配搭烏冬或素麵

★

椰菜
肉片鍋

Sliced Pork and Cabbage Hot Pot

Ingredients
材料

豚肉片 1 包	1 pack pork slices
椰菜 1/2 個	1/2 cabbage
芽菜 1 斤（600 克）	600 g mungbean sprouts
金菇 1 包	1 pack enoki mushrooms
芫荽 1 束	1 bundle coriander

Sauce
湯汁

鰹魚汁約 100 毫升
（按包裝濃縮比例調校）
柚子醋 2 湯匙
藤椒油 / 花椒油適量
薑蓉及蒜蓉各 1 湯匙
100ml bonito sauce
(dilute the sauce according to
the instruction)
2 tbsp yuzi vinegar
zanthoxylum pepper oil
or Sichuan peppercorn oil
1 tbsp minced ginger
1 tbsp minced garlic

Method
做法

1. 所有蔬菜洗淨，切好備用。
2. 豚肉片解凍，備用。
3. 鍋內先放油、薑蓉、蒜蓉炒香，加入湯汁及材料於火鍋爐內，蔬菜鋪滿鍋，放入豚肉片煮熟，以蔬菜伴吃。

1. Clean and prepare the vegetables.
2. Defrost the pork slices.
3. Stir-fry ginger and garlic until fragrant. Add the pork and the sauce in the pot, cook until done. Serve the pork with the vegetables.

Cold Tofu with Bonito Flakes

★

木魚碎冷豆腐

Ingredients
材料

豆腐 1 份　　　1 portion tofu
木魚碎適量　　bonito flakes
黑白芝麻適量　black and white sesames
紫菜碎適量　　seaweeds

Sauce
湯汁

鰹魚汁 1 湯匙（用開水調勻）
青檸汁 1/2 茶匙（可省略）

1 tbsp bonito sauce (dilute the sauce with drinking water)
1/2 tsp lime juice (optional)

Method
做法

1. 豆腐瀝乾水分，放於碟上。
2. 加入汁料，最後灑上木魚碎、黑白芝麻及紫菜碎即可。

1. Drain the tofu well. Place on the plate.
2. Add the sauce and sprinkle with bonito flakes, sesames and seaweeds. Serve.

Skye's tips

想增添飽肚感可以挑戰超級食物——納豆，建議大家將納豆配搭冷豆腐，絕對是倩揚最愛前菜之一。

Quick Summary

建議每日飲水量

體重kg X 40ML

你得我得行動組
Facebook 專頁

What to eat?

A 綠葉菜

B 其他蔬菜

C 蛋白質

D 原形食物澱粉質

水果　　　　蔬菜

全穀類 優質油脂
豆類及 堅果 適量運動

蛋類及
乳製品　　　肉/海鮮

✓ 每朝空腹上磅

✓ 記得飲夠水

✓ 整理食物記錄

✓ 入行動組 Check In

✓ 鼓勵自己：一定得！

Chapter 6

Hot Pot

打 邊 爐 飽 住 瘦
Recipes

自家製作健康低糖低鈉湯底，
配上新鮮蔬菜、低脂高蛋白肉類，
毋須捱餓，開開心心地飽住瘦打邊爐！
減肥可以食得飽，還可以約朋友！

　　食得健康與過生活，絕對可以取得平衡！看過前面的篇章，來到這裏應該不會覺得「減肥要捱餓」、「減肥無啖好食」、「減肥無得約朋友」了吧！如果還是想不到可以約朋友吃甚麼才符合減醣標準？打邊爐是一個 Gathering 的理想選擇！除了外出吃火鍋，也可以隨時在家打邊爐，只是跟着以下幾點，打邊爐飽住瘦，無！難！度！

1. 外食選清湯、昆布、豆乳等低鈉湯底，也可自家製少鹽少糖少油湯底。

2. 火鍋配料多選新鮮蔬菜、肉類及豆類製品。

3. 避免吃加工、油炸食物。

4. 選低脂、高蛋白的肉類。

5. 避用高鈉醬汁，多選低鹽豉油配天然食材，例如蒜蓉、辣椒、葱及薑等。

6. 火鍋進食次序，應先吃蔬菜打底，然後再吃肉類。

7. 最理想分為鴛鴦鍋，一邊灼肉類另一邊灼蔬菜，避免蔬菜多吸了肉類的油分而一併吃下。

8. 可選根莖類蔬菜如薯仔、南瓜及淮山等補充澱粉質，避免進食公仔麵及油麵等加工及油分較多的精製澱粉質。綠豆粉絲、米粉及蕎麥麵也是可接受的選擇。

韓式
牛排骨湯

(Korean Beef Short Rib Soup)
Galbitang

Ingredients 材料

牛排骨約 3 磅（1.35 公斤）	about 1.35kg beef short ribs
蘿蔔 1 條	1 radish (daikon)
洋蔥 1 個	1 onion
大蔥 1 條	1 sprig Peking scallion
蔥 1 條	1 sprig spring onion
蒜頭 8-10 瓣	8-10 cloves garlic
乾蔥 3 粒	3 shallots
薑 6 片	6 slices ginger
昆布乾 1/3 碗	1/3 bowl kombu strips (dried kelp)
魚乾 10 條	10 dried anchovies
雞蛋（隨個人分量）	eggs
綠豆粉絲適量	mungbean vermicelli (substitute for potato starch noodles)

Seasonings 調味料

韓式辣椒粉 （按個人接受程度）	Korean red pepper powder (depend on your preference)
生抽 2 湯匙	2 tbsp light soy sauce
麻油 2 茶匙	2 tsp sesame oil
鹽適量	salt
胡椒粉或黑椒碎適量	white pepper powder or chopped black pepper

打邊爐飽住瘦

Method
做法

1. 牛排骨洗淨，凍水下鍋飛水，沖洗乾淨。
2. 薑切片；蒜頭拍扁；乾葱切件。
3. 洋葱切長條；葱白及青葱切段；蘿蔔及大葱切滾刀塊。
4. 雞蛋拂勻，煎成薄蛋皮，切絲備用。
5. 綠豆粉絲用水浸軟，煮熟備用。
6. 油熱下鍋，加入薑片、蒜頭及乾葱炒香，加入牛排骨煎香兩面。下大葱、洋葱、蘿蔔、昆布乾及魚乾，加水蓋過面，加蓋，以中火燜約 40 分鐘（期間攪拌材料）。
7. 將蘿蔔及牛排骨取出，放於碟上。
8. 如想湯底清徹，可湯汁用隔油壺濾去多餘油脂，將牛骨清湯倒回鍋內，加蒜蓉、青葱及葱白煮 5 分鐘。
9. 最後可按個人喜好調味，伴綠豆粉絲或作為火鍋湯底。

1. Rinse the beef short ribs thoroughly. Place them in a pot of cold water and bring to a boil. Rinse the ribs again to remove any impurities (brown scums and fat).
2. Slice the ginger. Flatten the garlic cloves and cut the shallots into pieces.
3. Cut the onion into thick and long strips. Cut the white part and green part of the spring onion into sections. Cut the radish and Peking scallion into pieces diagionally.
4. Beat the eggs and make a thin omelet. Cut it into strips and set aside.
5. Soak the green bean vermicelli in water and cook it separately.
6. Stir-fry ginger slices, crushed garlic and shallots until fragrant. Add the beef short ribs and brown them on both sides. Put the scallion, onion, radish, dried kelp and dried anchovies to the pot. Add enough water to cover the ingredients. Cover with the lid and simmer over medium heat for about 40 minutes (occasionally stir the ingredients during the cooking process).

7. After 40 minutes, remove the radish and beef short ribs from the pot and transfer them to a separate plate.

8. If you prefer a clearer soup, strain the soup to remove excess fat using a strainer or fat separator. Pour the clear soup back into the pot. Add minced garlic, green and white part of spring onion. Let it simmer for 5 minutes.

9. To serve, add the soup base, beef short ribs, green bean vermicelli, and desired seasonings according to your personal taste.

Skye's tips

緊記放入打邊爐材料前，先品嘗一下牛骨湯的鮮甜味。平日也可做這個湯配搭小米或湯飯享用。

Tomato Potato Fish Soup

★

番茄薯仔
魚湯

Ingredients
材料

紅衫魚 2 份   2 portions red snapper
薯仔 2 個   2 potatoes
番茄 4-6 個   4-6 tomatoes
薑 5 片   5 slices ginger
粟米（可隨喜好加添）   the cob of corn (optional)
豆腐（可隨喜好加添）   tofu (optional)

Method
做法

1. 紅衫魚洗淨（包括魚肚），去鱗，細心檢查有否魚鈎，用廚房紙吸乾水分。
2. 薑洗淨外皮，切片備用。
3. 番茄洗淨，切件；薯仔去皮，切件備用。
4. 深鍋內油熱起鑊，放入薑片炒香，加入紅衫魚，加蓋（以防濺油），煎至兩面金黃，放入魚袋內。
5. 將火力調至最大，倒入滾水以大火煲約 20 分鐘或至魚湯呈奶白色，調至中火，加入番茄、薯仔煲 20 分鐘（或配粟米、豆腐等），熄火，加少許鹽調味，成為健康的火鍋湯底。

1. Clean the red snapper, make sure to remove any scales and clean the fish belly. Check carefully to ensure there are no fishing hook remaining. Pat dry the fish with the kitchen towels thoroughly.
2. Slice the ginger.
3. Clean the tomatoes and cut them into pieces. Peel and cut the potatoes into chunks.
4. Stir-fry the ginger slices until fragrant. Add the fish in the pot. Be prepared to immediately cover the pot with a lid to prevent oil splatters. Pan-fry the fish until both sides turn golden brown, then transfer them to a fish bag.
5. Turn to the high heat and pour in boiling water. Boil for about 20 minutes or until the fish soup turns milky white. Reduce to low heat and add tomatoes and potatoes (you can also add corn or tofu according to your preference). Simmer for 20 minutes, then season with a little salt before serving.

＊善用番茄的鮮甜及水分，煲湯時多加番茄，少一點水分，湯底更濃郁美味！

Skye's tips

· 如沒有魚袋，記得盛湯時用隔篩過濾一次，以防魚骨。
· 如怕魚袋有破損，倒進火鍋爐時可用隔篩過濾一次，更安心！

打邊爐飽住瘦

★ 倩揚牌
秘製湯底

Ingredients
材料

櫻花蝦 1/3 碗	1/3 bowl dried sakura shrimps
乾秀珍菇 1/3 碗（用水泡浸）	1/3 bowl dried oyster mushrooms
	(soak in the water until soft)
大葱 1 條	1 sprig Peking scallion
葱 2 束（分葱白及青葱）	2 bundles spring onion
	(separate white part and green part)
乾葱 3 粒	3 shallots
蒜頭 10 瓣（拍扁）	10 cloves garlic (pat lightly)
薑 5 片	5 slices ginger
滷水汁適量（製法參考 p.137）	braising sauce (refer to p.137)

Seasonings
調味料

＊按口味自行調味	***adjust the taste according to your preference**
味醂	mirin
料理酒	cooking wine
鰹魚汁	bonito sauce (hondashi)

1. 油熱起鑊，下薑片炒香，放入乾葱、大葱及葱白炒勻，加入蒜頭、櫻花蝦、乾秀珍菇及青葱一同炒香。
2. 倒入滷水汁，加水至湯煲的 3/4 深度，拌入調味料煮滾。

1. Stir-fry ginger until fragrant. Put in shallot, Peking scallion, white part of spring onion. Mix well. Add the garlic, dried sakura shrimps, dried oyster mushrooms and green part of spring onion and stir-fry for a while.
2. Pour in the braising sauce, add the water to fill the pot to about 3/4 of its depth. Add the seasonings and bring to boil.

Skye's tips

薑、葱、蒜等料頭一定要準備充足，以帶起整個湯底的香氣，亦可減少含鹽分的調味料，例如可少下一點鹽，滷水汁亦可減少。另外，建議湯底滾好後先取起薑、葱及蒜等食材，有時我也會加一大把芫茜或其他蔬菜先灼，先吃蔬菜打底也是打邊爐的食住瘦秘訣啊！

打邊爐飽住瘦

飽住瘦打邊爐，健康食材精選

深綠色蔬菜及菇類

西洋菜、皇帝菜、油麥菜、萵筍、本菇、蘑菇、金菇

優質澱粉質

番薯、蓮藕、南瓜、薯仔、粟米、淮山

蛋白質及纖維

豚肉片、雞件、蝦、蟶子、鮑魚、魚片、花蛤、豆腐、芋絲

輕鬆減，飽住瘦，無難度！
健康減醣生活由這刻開始！

看過以上倩揚的食譜餐單，相信大家都認識了更多日常減醣適用的健康食材，並懂得善用天然調味料預備省時常備菜，以及掌握了如何烹調「減醣餐不用分開煮」的一日三餐。

每天實行減醣生活的基本原則：

1. 多元化的食物選擇，攝取均衡營養。
2. 每日進食大量蔬果，包括兩份拳頭大的水果。
3. 攝取優質的高蛋白、澱粉質及油脂。
4. 多吃全穀類、豆類及堅果。
5. 每天飲用大量清水。
6. 勤做運動。

由今日開始好好經營健康！
向多餘脂肪講拜拜！
希望大家擁有健康的身體！
努力堅持！加油！

陳倩揚 低醣飲食 系列書籍

豐富你的減醣知識，與你並肩同行！

1 輕鬆減磅概念入門 家常減醣餐單分享

2 更豐富減醣營養全書 加強第一本概念實踐

3 「倩揚廚房」減醣食譜 輕鬆備餐 Let's Go!

念妍方 黑糖四物膏

氣血補給 妳要好「物」友

十多種中草藥成分提煉
有助緩解減肥氣血不足

京都念慈菴

念妍方
黑糖四物膏

減肥易洩「氣」?
減到脫髮·M痛·體虛

購買詳情

f Feminin.hk   Feminin.hongkong

念妍方黑糖四物膏 ✕ 倩揚補氣血食譜

黑糖四物滷味
(可自行加入喜歡的蔬菜或肉類)

更多補氣血食譜

材料：　　　2-4人份

硬豆腐	150g/1盒
蒟蒻	250g/1包
蛋	4隻
乾海帶芽	1/2碗
雞鎚	10隻
椰菜花	1碗
粟米芯	8條
杏鮑菇	2條

調味料：

念妍方黑糖四物膏	2包		
醬油膏	5湯匙	蔥	2束
味醂	3湯匙	辣椒	4條
料理酒	3湯匙	薑	6片
老抽	2湯匙	蒜	4粒
水	適量		
滷包	1包		
大蔥	2條		

做法：

① 首先加入適量食油，將大蔥、薑、蔥及蒜爆香。

② 然後加入醬油、味醂、水、滷包。

③ 煮滾後，先加入肉類煮10分鐘(肉類可預先汆水)，蛋先煮熟，剝殼備用。

④ 然後加入其他素菜，用小火煮10分鐘，加蛋，熄火，加入2包念妍方黑糖四物膏攪拌均勻，蓋上鍋蓋，浸1小時就可食用。

至Pro媽咪
紙選Pro-X廚紙

特吸X型壓紋
3層
吸水
+40%^

全新 **維達Pro-X**
特吸萬用廚紙

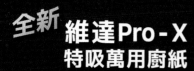

3層強韌厚實　特吸鎖水X型壓紋　3重食品級安全認證#

4卷裝　　抽取式

 VINDA HONG KONG | Q　 vinda_hongkong | Q　　　**各大超市有售**

低醣飲食生活提案 **3**
健康瘦身餐
LOW-CARB MEAL PREP

著者
陳倩揚

責任編輯
簡詠怡

裝幀設計、排版
羅美齡

封面及插圖設計
陳倩揚

封面攝影
Alien Creation Limited

攝影
梁細權

出版者
萬里機構出版有限公司
香港北角英皇道 499 號北角工業大廈 20 樓
電話：2564 7511　　傳真：2565 5539
電郵：info@wanlibk.com
網址：http://www.wanlibk.com
　　　http://www.facebook.com/wanlibk

發行者
香港聯合書刊物流有限公司
香港荃灣德士古道 220-248 號荃灣工業中心 16 樓
電話：2150 2100　　傳真：2407 3062
電郵：info@suplogistics.com.hk
網址：http://www.suplogistics.com.hk

承印者
美雅印刷製本有限公司
香港觀塘榮業街 6 號海濱工業大廈 4 樓 A 室

出版日期
二〇二三年七月第一次印刷
二〇二三年九月第二次印刷

規格
16 開（170 mm × 230 mm）

鳴謝場地提供
煤氣烹飪中心